GUIDE TO BRICK, CONCRETE, AND STONEWORK

McGraw-Hill Paperbacks
Home Improvement Series

Guide to Plumbing

Guide to Electrical Installation and Repair

Guide to Roof and Gutter Installation and Repair

Guide to Wallpaper and Paint

Guide to Paneling and Wallboard

Guide to Landscape and Lawn Care

Guide to Brick, Concrete, and Stonework

Guide to Carpentry

Guide to Furniture Refinishing and Antiquing

Guide to Bathroom and Kitchen Remodeling

GUIDE TO BRICK, CONCRETE, AND STONEWORK

McGRAW-HILL BOOK COMPANY

New York   St. Louis   San Francisco   Auckland   Bogotá   Düsseldorf
Johannesburg   London   Madrid   Mexico   Montreal   New Delhi   Panama
Paris   São Paulo   Singapore   Sydney   Tokyo   Toronto

1 2 3 4 5 6 7 8 9 0 SMSM 8 3 2 1 0

Library of Congress Cataloging in Publication Data

Main entry under title:

Guide to brick, concrete, and stonework:

  (McGraw-Hill paperbacks home improvement series)
  Originally published in 1975 by Minnesota Mining and Manufacturing
Company, Automotive-Hardware Trades Division, St. Paul, in the Home
pro guide series, under title: The home pro brick, concrete, and stonework
guide.
  1. Masonry-Amateurs' manuals. I. Minnesota Mining and
Manufacturing Company. Automotive-Hardware Trades Division. Home
pro brick, concrete, and stonework guide.
TH5313.G84   1980   693'.1   79-26982
ISBN 0-07-045967-3

Cover photo: Findlay, Kohler Interiors
            Columbus, Ohio

            Labrenz, Brown and Reimer
            Landscape Architects
            Columbus, Ohio

Contents

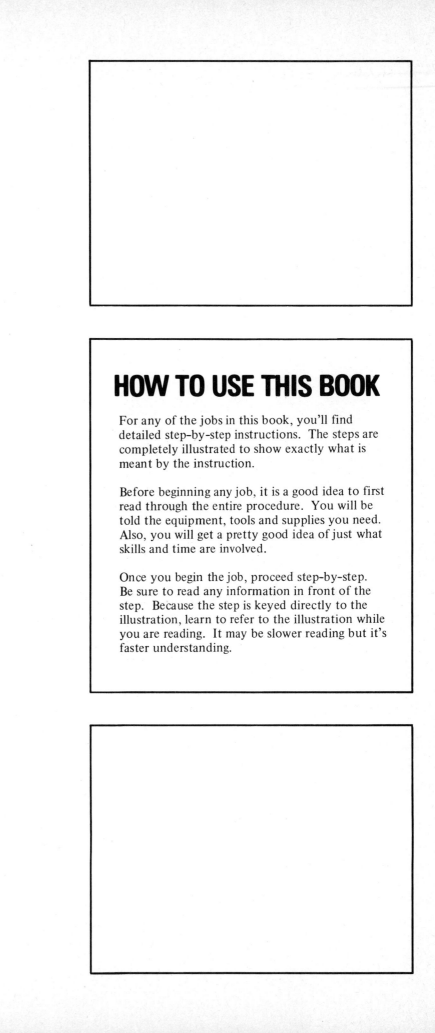

HOW TO USE THIS BOOK

For any of the jobs in this book, you'll find detailed step-by-step instructions. The steps are completely illustrated to show exactly what is meant by the instruction.

Before beginning any job, it is a good idea to first read through the entire procedure. You will be told the equipment, tools and supplies you need. Also, you will get a pretty good idea of just what skills and time are involved.

Once you begin the job, proceed step-by-step. Be sure to read any information in front of the step. Because the step is keyed directly to the illustration, learn to refer to the illustration while you are reading. It may be slower reading but it's faster understanding.

GUIDE TO BRICK, CONCRETE, AND STONEWORK

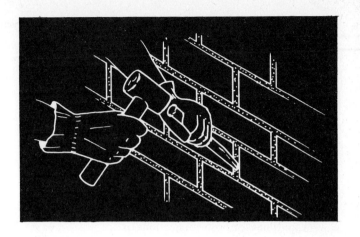

BRICKWORK

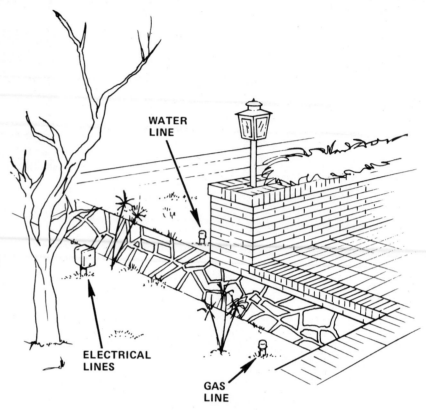

▶ Descriptions and Uses

When planning your brick addition, consider future architectural or landscape changes. Make provisions for these future projects at this time — even if they will not come for some time.

Some typical future changes might be:

- Water lines for hose connections or sprinkler systems.

- Electrical lines for future utilities or lighting.

- Gas lines for future outdoor barbeque, pool heater, fire pit.

Consider the future growth of landscape vegetation and their root systems. Possibly you will want to remove or relocate a tree or shrub.

▶ Building Codes

The first step in making brick additions is to become familiar with:

- The building codes and construction ordinances in your area.

- The minimum property or construction requirements of the Federal Housing Authority (FHA) and the Veterans Administration (VA).

Your brick addition must comply with the minimum standards of the FHA and VA requirements if these agencies are to guarantee future mortgage loans on the property.

Various agencies are responsible for building regulations in large cities. Your local City Hall can direct you to the local agencies that maintain the zoning ordinances for your property and the building codes for your kind of brick addition.

These agencies issue building permits, if permits are required for your brick addition. They furnish inspectors to verify that the brick addition conforms to the specification requirements.

Some regulations affect brick additions from the durability and safety standpoint. Other regulations affect brick additions in that zoning ordinances may restrict certain architectural features.

The agencies also maintain published standard specifications that govern construction. These publications can be a big help to the home pro. They can be an excellent source of information and provide specific design recommendations.

The instructions in this section do not necessarily conform to all the codes and regulations required for your particular area.

▶ Brick Paving

Brick paving must be laid on a stable surface. There are two basic practices that are commonly accepted:

- Laying bricks in a bed of mortar over a concrete slab. The slab thickness and the use of reinforcements make this the stronger and more permanent construction.

- Laying bricks in a bed of sand at least 2 inches thick.

When laying bricks on sand, there are three generally accepted methods for adding to the stability of the paving.

1. Lay the bricks as close together as possible and fill any gaps with sand.

2. Space the bricks 1/8- to 3/8-inch apart and fill the gaps with dry mortar mix. When the surface is wetted, the mortar will harden.

3. Space the bricks 3/8- to 1/2-inch apart and fill the gaps with mortar.

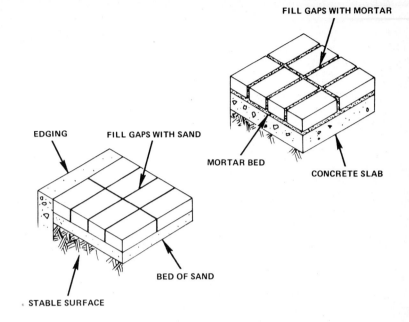

The edges of brick pavement merit special consideration. Whether the bricks are to be laid in a bed of sand or a bed of mortar, it is essential that the brick pavement has edging so that the bricks do not spread out. Concrete, brick or wood can be used for edging.

Brick Paving

The requirements of edges and dividers must be considered when planning and designing your paving project.

Concrete edges [1] and dividers are narrow slabs that:

- Must be at least 4 inches thick.

- Must be made on a proper base. Refer to the POURED CONCRETE section for specific information on concrete slabs.

Brick edges and dividers:

- May be laid face up [2], side up [5], end up [4] or turned in any direction.

- Must be set in a bed of mortar over a concrete slab [3] at least as wide as the edging. The slab must be at least 3 inches thick.

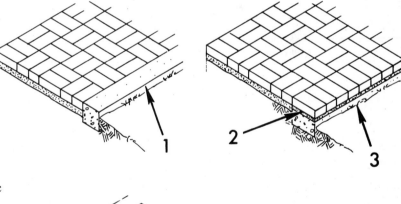

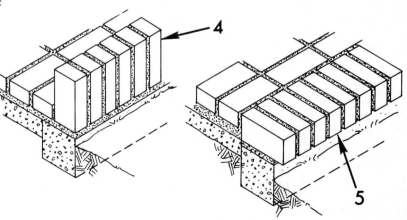

DESIGN CONSIDERATIONS

Brick Paving

Wooden edges and dividers:

- Should be made of materials that are resistant to weathering, rotting and insects. Redwood or locust are excellent choices.

- Should be treated with a commercially available preservative to lengthen the life of the material.

- Must be carefully aligned and leveled.

- May be rigidly nailed together.

- May be embedded in concrete. If not embedded in concrete, edging must be firmly staked and supported at frequent intervals on a stable surface such as rocks or broken bricks.

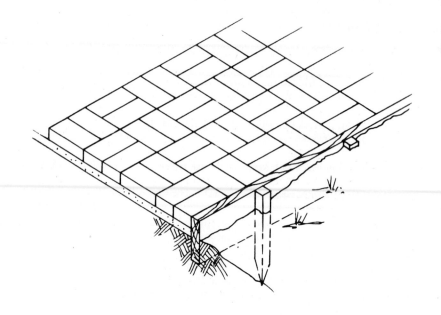

▶ **Driveways**

When planning your driveway, the following requirements must be considered. The specific criteria may not necessarily agree with local codes and regulations.

If no local codes or regulations exist, refer to minimum standards of the FHA or VA requirements.

- Entrance should have a flare or radii for safe and convenient entrance and exit.

- Structural features must prevent damage from freezing or flash floods.

- Bricks should be mortared in place on a concrete slab at least 4 inches thick. Refer to the POURED CONCRETE section for specific information on concrete slabs.

- Surface should slope to permit adequate water runoff. Slope should be designed to prevent the bottom of the vehicle or its bumper from hitting the surface.

  If possible, limit the slope to no more than 1-3/4 inch per foot.

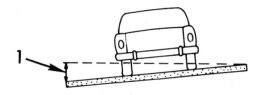

If the driveway is essentially level, provide a slope across the driveway. The cross slope [1] should be 1/8-inch per foot minimum, 5/8-inch per foot maximum.

- The driveway should be wide enough to permit pedestrian traffic past a parked vehicle and permit loading and unloading of vehicles while standing on the pavement.

  The recommended width of a driveway is 10 feet per vehicle. The minimum width should be no less than 8 feet per vehicle.

  Curved sections must be wider than straight sections to allow vehicles to turn.

▶ Sidewalks

When planning your sidewalk, be sure to check local codes and regulations and minimum standards of the FHA or VA requirements.

The following requirements must be considered when planning your sidewalk. Specific criteria may not necessarily agree with local or FHA/VA requirements.

- The recommended minimum width of access or service sidewalks is 2 to 3 feet. The minimum width of a public sidewalk or one leading to the entrance of your home should be 4 feet.

- Surface should slope to permit adequate water runoff. Slope should be away from buildings.

  The slope of a sidewalk must not exceed 5/8-inch per foot if area is subject to freezing, or 1-3/4 inches per foot if area is not subject to freezing. If the slope exceeds these limits, the sidewalks will not be safe to use during bad weather. If the slope must exceed these limits, consider using steps.

If the sidewalk is essentially level, provide a slope across the sidewalk. Cross slope [1] should be 1/8-inch per foot minimum, 5/8-inch per foot maximum.

- Bricks may be laid in a bed of sand, or mortared in place on a concrete slab at least 3 inches thick.

  Refer to the POURED CONCRETE section for specific information on concrete slabs.

  If bricks are to be laid in a bed of sand, use a bed of sand at least 2 inches thick on well-drained soil. On poorly drained soil, correct the drainage problems. Then use a bed of sand at least 2 inches thick.

▶ Patios

When planning your brick patio, the following requirements must be considered. The specific criteria may not necessarily agree with local codes and regulations.

If no local codes or regulations exist, refer to the minimum standards of the FHA or VA requirements.

- Surface should slope to permit adequate water runoff. Slope should be away from buildings or toward a drain.

  The minimum recommended slope is 1/8-inch per foot.

- Bricks may be laid in a bed of sand or mortared in place on a concrete slab at least 3 inches thick.

  Refer to the POURED CONCRETE section for specific information on concrete slabs.

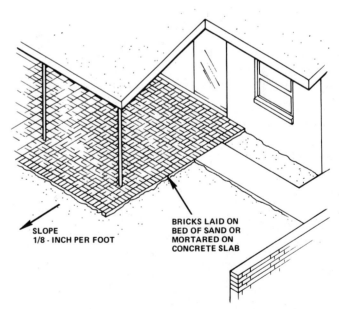

SLOPE 1/8 - INCH PER FOOT

BRICKS LAID ON BED OF SAND OR MORTARED ON CONCRETE SLAB

If bricks are to be laid in a bed of sand, use a bed of sand at least 2 inches thick on well-drained soil. On poorly drained soil, correct the drainage problems. Then use a bed of sand at least 2 inches thick.

DESIGN CONSIDERATIONS

▶ **Brick Walls**

When planning your brick wall, the following requirements must be considered. The specific criteria may not necessarily agree with local codes and regulations.

If no local codes or regulations exist, refer to the minimum standards of the FHA or VA requirements.

- Decorative walls may not be regulated. Check local ordinances.

- Retaining walls must meet more rigid structural requirements than decorative walls.

  In general, all retaining walls more than a few feet high and all structural masonry walls can be expected to be governed by building codes and/or FHA/VA requirements.

- Brick walls must be made on footings or foundation slabs. Refer to the POURED CONCRETE section for specific information on footings and slabs.

Proper footing or foundation is important, especially in heavy soil that absorbs and retains lots of water, or in areas where drainage is a problem.

A footing is an enlargement at the base of a wall that distributes the weight of the wall on the soil.

The footing should always have its base below the frost line.

Brick Walls

The footing [1] should be at least as thick as the wall is thick.

The footing should be twice as wide as the thickness of the wall. If the soil is unstable, the footing should be more than twice as wide as the wall is thick.

- Footing for wall should be horizontally level.

- The bricks that rest on the footing, or bricks that are below the surface of the ground, must be the severe weathering (SW) grade or concrete blocks. They will be exposed to the surface moisture. If they absorb water, the water may freeze and the bricks may break.

- Consider the environment to which the wall will be exposed, and select the type of brick that is designed for that kind of environment. Refer to information on Page 11 (bottom) about grades of brick and their uses.

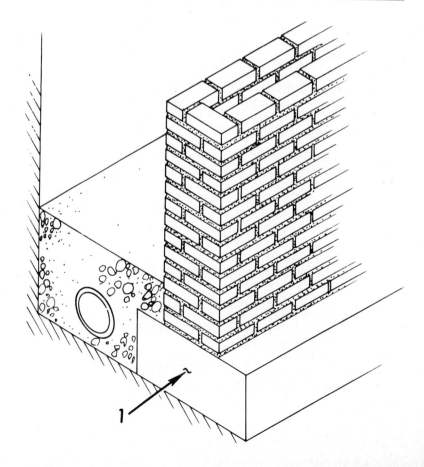

Brick Walls

- If reinforcements are required in the wall, as prescribed by code or regulation, provision must be made for reinforcements before the footing is made. Vertical reinforcement rods are spaced in accordance with building code regulations.

  Vertical reinforcement rods can be one piece or two pieces.

If one-piece vertical reinforcement rods [1] are used:

- The bent end is generally 12 times the diameter of the rod.

- The radius of the bend should be 7 times the diameter of the rod.

- The horizontal section should be at approximately the center of the footing.

- The top of the reinforcing rod should be 1/2 brick short of reaching the top of the wall.

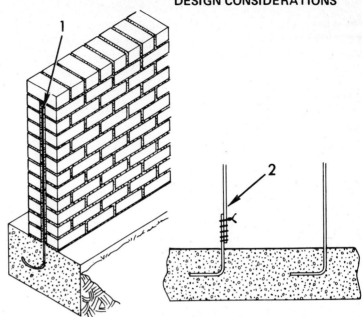

If two-piece vertical reinforcement rods [2] are used, the vertical sections are tied to the footing sections. The above criteria applies to the height of the rods and requirements of the bent end. Use corrosion-resistant wire to tie the two rods together.

▶ **Steps**

Brick steps may be made using the same general practices as applied to brick walls.

Steps can be made in such a wide selection of designs that it is not practical to give specific guidelines to cover all of the possibilities.

There are some guidelines that may be helpful, however.

- The design of your brick steps should comply with the minimum standards of the FHA or VA requirements. Steps must comply with the requirements of local building codes.

- Steps should have uniform tread depth.

- Tread should be about 1/4-inch higher at the rear than at the front to permit water to run off.

- Step risers should be uniform in height. Some building codes specify limits on height of risers.

- Risers must be level from side to side.

- Landings must be built at appropriate intervals. The width of the landings must be adequate.

- Provisions must be made for anchoring railings if applicable.

- Steps must extend beyond the door sill.

- If the outer edge of a step is cantilevered, the bricks should be placed endwise to provide the most leverage for the mortar.

- Expansion joints must be built between step masonry and other masonry. Refer to the POURED CONCRETE section for specific information on expansion joints.

DESIGN CONSIDERATIONS

Steps

- To prevent steps from settling or moving away from adjacent structure, footings are usually required.

  Footings for steps may be made of concrete or bricks.

  If footing is made of concrete, refer to the POURED CONCRETE section for specific information on footings and slabs.

  If the footing is made of brick, severe weathering (SW) grade bricks should be used and the footing should be plastered with grout. Refer to the CONCRETE BLOCK section for specific information on parging.

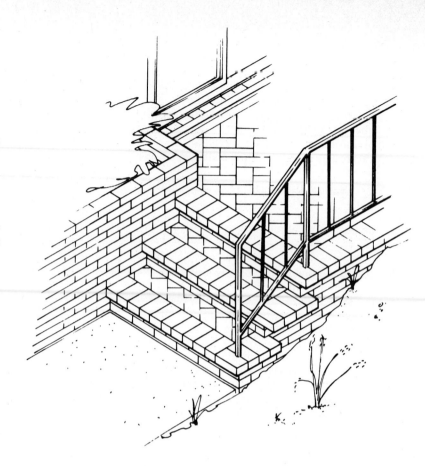

▶ Outdoor Fireplaces

When planning your brick fireplace, the following requirements must be considered. The specific criteria may not necessarily agree with local codes and regulations.

If no local codes or regulations exist, refer to the minimum standards of the FHA or VA requirements.

- Brick fireplaces must be made on footings or foundation slabs. Refer to the POURED CONCRETE section for specific information on footings and slabs.

  If reinforcements are required, the rods must be positioned carefully.

- A clay flue liner is recommended. It should be supported at the bottom on at least three sides by projecting bricks.

  The flue liner [1] should project approximately 4 inches above the masonry. The masonry can be topped with mortar.

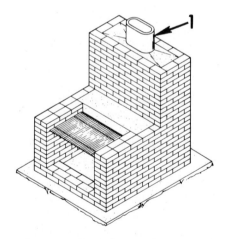

The flue should be installed using a cement mortar, which is more heat-resistant than lime mortar.

A good mortar mix is:

  1 part Portland cement
  1 part lime
  6 parts clean sand

Mix ingredients by volume.

Outdoor Fireplaces

- The fire pit of a fireplace should be lined with fire brick. Other kinds of bricks can not withstand the temperatures and will crack or crumble. Fire brick should be laid using fire clay mortar with very thin mortar joints.

- A commercially available grill may be purchased and installed according to the manufacturer's recommendations
or
A grill may be mounted in the fireplace by embedding 3/8-inch rods in the mortar.

- A steel plate between the grill and the chimney may be handy for cooking.

- Grill and steel plate should be removable to permit easy cleaning of the cooking surfaces and the fire pit.

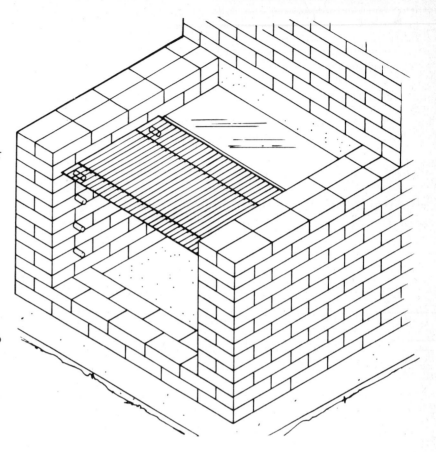

▶ Brick Veneers

Plain monotonous surfaces are often covered with a brick veneer to provide a more interesting or more durable surface.

Visit your masonry supplier to view the selection of veneering materials available.

Although brick veneers are usually not structural members, it is recommended that you check local codes and regulations before beginning work on a veneer addition.

There are two basic types of veneer:

- Simulated plastic or ceramic bricks that are installed over existing surface using adhesives or mortar.

    Simulated bricks are thin, easy to handle and install. They are available in large panels or individual bricks.

- Genuine bricks that are used to build a brick wall parallel to the existing surface. The parallel wall is called a brick veneer.

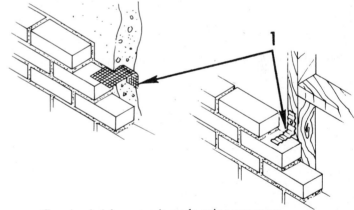

Genuine brick veneer is made using common bricks and mortar. They must be installed on a solid base. See Page 18.

The veneer wall should be anchored to the structure using bonding strips [1].

Bonding strips can be wire fabric embedded in the mortar bed of the brick veneer.

Bonding strips may be embedded in the mortar of masonry structure or nailed to the studs of a wooden structure of the structural wall.

▶ Tools

The following tools are required to make brick walls or paving:

- A wheelbarrow [1].

- A cement mixer [2], if desired, to mix mortar in large quantities.

- Mortar box [3] to mix mortar in small or moderate quantities.

- A brick trowel [4] to spread mortar. Trowel should have a durable handle to tap bricks into final position in the mortar. Buy a brick trowel with a comfortable length blade. If the blade is too long, your wrist will become tired easily.

- A pointing trowel [5]. This is a small version of the brick trowel. It is generally used for repairing or tuck-pointing mortar joints. It is not used to spread large amounts of mortar.

- A rule [6].

- A bucket [7] and brush [8] for cleaning up brick surfaces.

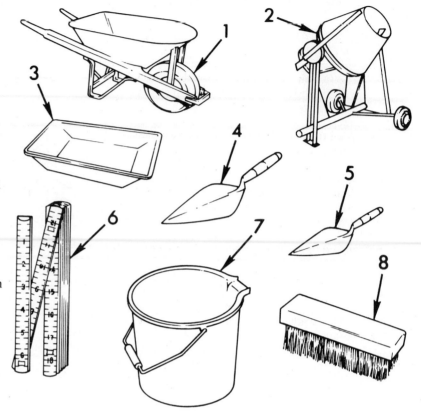

Tools

- A shovel [1].

- A bricklayer's set [2] or mason's chisel [3]. They are used with a brick hammer [4] to cut and trim bricks.

- A mason's level [5] to check that bricks in walls are level and plumb. Also used to check the slope or grade of brick pavements.

- A mason's line [6] to mark the top level position of bricks in the courses. A course is a row of bricks. Any stout cord can serve as a mason's line.

- A story pole [7]. This is a board marked at intervals corresponding to the height of each course in a brick wall. It is used to accurately check the height of any point of a wall.

- Corner blocks [8]. They are used to hold the mason's line at each end of the course.

- A line level [9] to check that the mason's line is horizontally level.

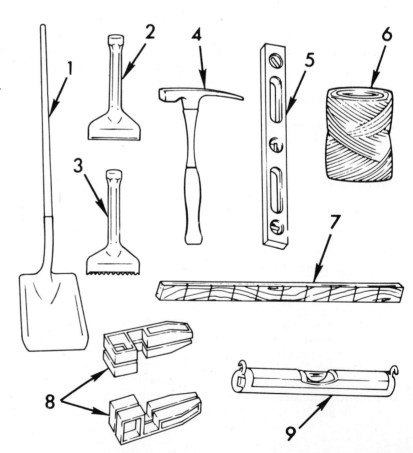

Tools

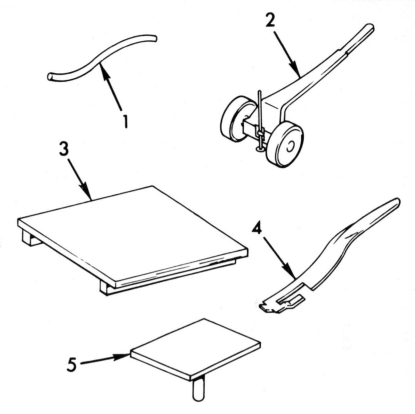

- A brick jointer [1] to compact the mortar into the mortar joints. Compacting makes mortar joints weatherproof and uniform in appearance.

- A joint raker [2] to remove mortar from mortar joints to a desired depth.

- A mortar board [3] to transport small quantities of mortar from the mortar mixing site to the place where the mortar is used.

- A raker jointer [4] is a raker on one end and a jointer on the other.

- A hawk [5]. This is a small portable mortar board. It is especially useful when repairing mortar joints.

▶ **Supplies**

Mortar. Mortar is used as a bonding agent for masonry materials. Mortar is a mixture of Portland cement, hydrated lime, damp loose mortar sand and water. Different types of mortar are used for different purposes.

Bricks. The standard bricks used for brick additions are called common or building bricks. They come in three grades:

SW – Bricks intended for exposure to severe weather, including the possibility of being frozen when soaked with water.

MW – Bricks intended for exposure to severe weather, but not the possibility of being frozen when soaked with water.

NW – Bricks not intended for exposure to severe weather. Suitable for use with interior masonry.

Bricks for exterior use are usually SW or MW grade bricks which combine strength, durability and pleasing appearance.

Bricks come in many colors, textures and surface finishes. Colors vary so much that it is generally preferable to buy bricks to match a sample brick than to state a color.

Sometimes bricks are available with core holes. These bricks are lighter to use and provide a place for mortar to get a good bond.

Used bricks are very popular for landscaping. Manufacturers have recognized this popularity and produce "used bricks" to supplement the natural supply.

Although all sizes of bricks may not be available locally, bricks are made in the following sizes:

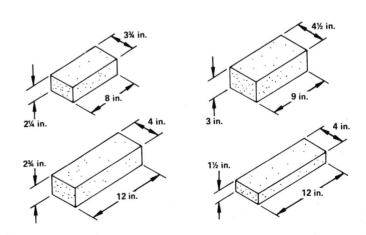

Material needs will vary with the size of bricks selected and the thickness of the mortar joint used. Material needs are also affected by variation in the use of bonding courses. In bonding courses, the ends of the bricks rather than the sides of the bricks face the wall.

Also, you will need to allow for some breakage and some waste mortar.

Table 1 presents the numbers of common bricks required for various wall areas if the sides of the bricks face the wall and every sixth course is a bonding course. The number of bricks required and the amount of mortar required have been increased by 5% to allow for breakage and waste.

Table 2 presents the numbers of common bricks required for various paved areas if the top of the brick is used to face the paved surface. The number of bricks required and the amount of mortar required have been increased by 5% to allow for breakage and waste.

The amount of mortar is based on using 1/2-inch mortar joints.

The tables apply for common U.S. bricks only.

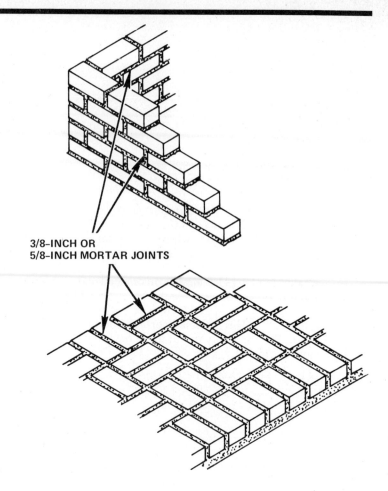

3/8-INCH OR 5/8-INCH MORTAR JOINTS

Table 1

Material Needs for Brick Walls

Wall Area in Square feet	For Walls with 1/2-Inch Thick Mortar Joints* Quantities Assume 5% Allowance for Waste					
	Wall Thickness in Inches					
	4 Inches		8 Inches		12 Inches	
	Number of Bricks	Cu. ft. of Mortar	Number of Bricks	Cu. ft. of Mortar	Number of Bricks	Cu. ft. of Mortar
100	756	8.5	1,512	17.0	2,268	25.5
200	1,512	17.0	3,024	34.0	4,536	51.0
500	3,780	42.5	7,560	85.0	11,340	127.5
1000	7,560	85.0	15,120	170.0	22,680	255

*For other mortar joint thicknesses, go to next page to estimate.

Table 2

Material Needs for Brick Pavement

Area in Square feet	For Pavement with 1/2-Inch Thick Mortar Joints* Quantities Assume 5% Allowance for Waste	
	Number of Bricks	Cubic feet of Mortar
100	419	3.2
200	838	6.4
500	2,095	16.0
1000	4,190	32.0

*For other mortar joint thicknesses, go to next page to estimate.

▶ **Walls**

- If 3/8-inch mortar joint is used, decrease the amount of mortar materials by 20% and increase the number of bricks by 6%. Refer to Table 1.

- If 5/8-inch mortar joint is used, increase the amount of mortar materials by 18% and decrease the number of bricks by 6%. Refer to Table 1.

▶ **Paving**

- If 3/8-inch mortar joint is used, decrease the amount of mortar materials by 22% and increase the number of bricks by 5%. Refer to Table 2.

- If 5/8-inch mortar joints are used, increase the amount of mortar materials by 19% and decrease the number of bricks by 5%. Refer to Table 2.

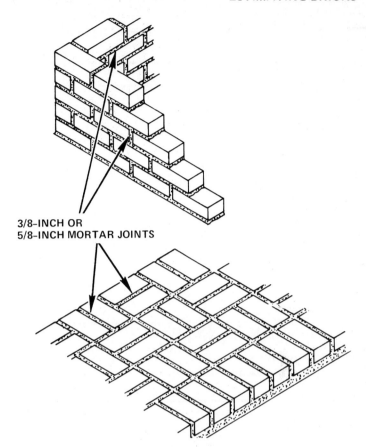

3/8-INCH OR
5/8-INCH MORTAR JOINTS

MIXING MORTAR

Mortar is made by mixing ingredients by volume. The proportions vary for different kinds of masonry.

Some accepted mixing proportions are listed in Table 4. To find the correct mixture for your job:
- Select the type of mortar required for your application from Table 3, **Page 14.**

- Table 4, **Page 14,** prescribes the proper proportions of various ingredients for the different types of mortar.

Mortar must be used within 2-1/2 hours (3-1/2 hours in cool weather) after initial mixing. Mortar not used within this time period should be discarded. Only mix the amount of mortar you will be able to use within the time limits.

Within the time limits, water may be mixed with the mortar to replace water lost by evaporation. This is called retempering.

Up to 15% coloring agents (3% lamp black or carbon black) may be safely added to mortar mix by volume.

If colored mortar is used, do not add water and remix the mortar. Retempering colored mortar may result in changing the color.

Normal mortar mixes will not withstand the temperatures in a brick fireplace. Special mortar is available for use with fire brick. Check with a masonry supply dealer for the correct mortar.

Table 3. Mortar Uses

Masonry Item	Type of Mortar Required
Structural Walls	M
Exterior Walls Above Grade	N
Exterior Cavity Walls	M
Reinforced Masonry	M
Interior Non-Bearing Walls	N
Retaining Walls	M
Tuck-Pointing (Repairs)*	M or N
*See Brickwork Repairs for special mixing instructions	

Table 4. Mortar Mix by Volume

Mortar Type	Portland Cement	Hydrated Lime or Lime Putty	Type II Masonry Cement	Damp Loose Mortar Sand
M — High Strength Mortar	1	1/4	—	3 to 3¾
	1	—	1	4½ to 6
N — Medium Strength Mortar	1	1/2 to 1	—	4 to 6
	—		1	2-¼ to 3

1. Combine the dry ingredients determined from table above in a mixing container. Thoroughly mix dry ingredients together until they are mixed evenly.

The proper water content is important to permit mortar to be easily applied to the bricks and to have a good bonding strength. If mortar is too wet, it will not retain shape or position on tools or bricks. If mortar is too dry, it will not adhere properly to the bricks.

The exact amount of water to use cannot be specified. As a rule, add as much water as you can without impairing the workability of the mortar. Mortar has good workability if it:

● Spreads easily.

● Clings to vertical surfaces and the underside of horizontal surfaces.

● Does not flow out of the joints between the bricks.

Use drinking-quality water only for best results.

2. Add water and continue mixing ingredients together until mortar has good workability. Mix ingredients for at least 3 minutes.

Mix or stir mortar frequently while using to keep mixture in proper consistency.

Mortar that falls to the scaffold or the ground should not be used. It should be discarded.

To ensure that your brick paving will be sound and durable, observe the following rules:

- Be sure that the area is properly graded and drained.

- Make proper bases for footings and slabs.

- Do not mix or work with mortar in cold or freezing weather.

- Be sure that bricks are properly wetted before applying mortar. The damp surface assists the spreading and bonding of the mortar. Because wet bricks do not absorb moisture out of the mortar, the mortar can cure properly. Bricks should be soaked for at least 1 minute before mortar is applied.

The first step in making brick paving is digging and grading the site. Then make bases, footings or concrete slabs as required.

Refer to the POURED CONCRETE section for specific information about:

- Grading and building bases and footings

- Making concrete forms

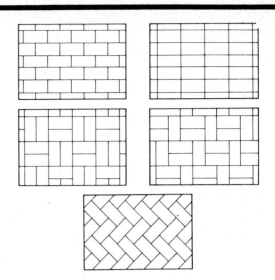

- Sidewalks and driveways

- Slabs and patios.

After the above work is completed, brick paving is made as follows:

If laying bricks in a bed of mortar over a concrete slab, go to next section (below).

If laying bricks in a bed of sand, go to Page 16.

▶ **Laying Bricks on Mortar**

1. Lay a bed of mortar [1] 5/8- to 3/4-inch thick on a portion of the concrete slab.

When laying bricks in the mortar bed, be sure that:

- Each brick is level and even with the other bricks.

- Mortar joints [2] between the bricks are not too wide. If joint is too wide, the pattern will not be continuous.

2. Lay bricks in the mortar bed. Press bricks firmly into the mortar.

When filling mortar joints, fill each joint a little higher than the brick surface. Mortar will shrink as it cures.

3. After laying several bricks, fill the mortar joints with mortar. Tamp mortar into the joints.

4. After mortar starts to harden, trim the mortar so that mortar is even with brick surface.

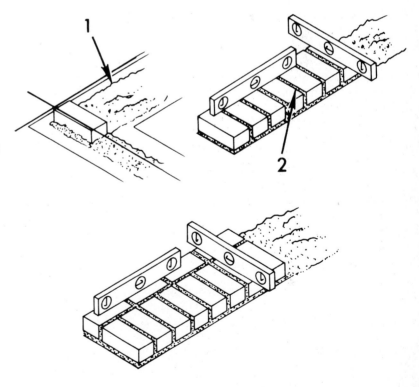

MAKING BRICK PAVING

▶ **Laying Bricks on Sand**

Be sure edges and dividers are properly installed.

1. Spread sand between edges and dividers to approximately the desired level.

2. Using a screed [1], level the sand to a height of 1/4- to 3/8-inch above the desired base level. Add or remove sand as required.

3. Using a fine spray of water, thoroughly wet the sand. Do not disturb the smooth surface.

4. Allow sand to dry.

Repeat Steps 1 through 4 three times. Add sand as required to maintain the desired level.

5. Lay bricks on smooth surface. Do not allow bed to be disturbed before laying the bricks.

As bricks are laid, be sure that each brick is level and even with the other bricks and that the gaps between the bricks are as small as possible.

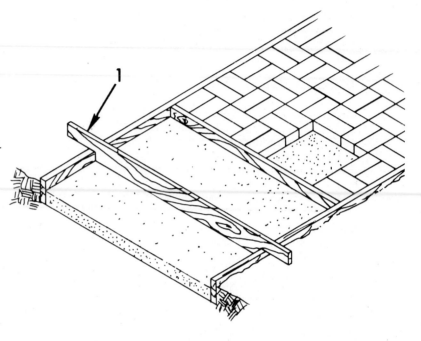

Laying Bricks on Sand

When making brick paving, there are three generally accepted methods for adding to the stability of the pavement:

● Using Mortar, below.

● Using Dry Mortar Mix, Page 17.

● Using Sand, Page 17.

Using Mortar

When filling the gaps [1] with mortar, fill each gap a little higher than the brick surface. The mortar will shrink as it cures.

1. Fill the gaps between the bricks with mortar. Tamp the mortar into the gaps.

2. After the mortar starts to harden, trim the mortar so that the mortar is even with the surface.

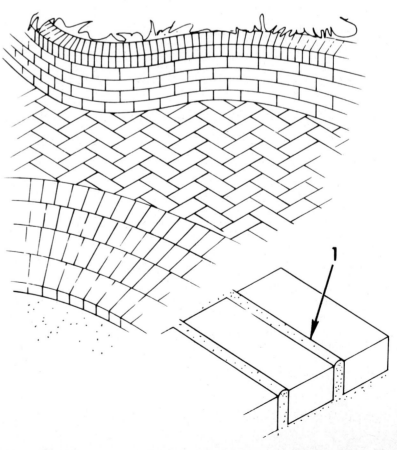

Laying Bricks on Sand

Using Dry Mortar Mix

Using dry mortar mix to lay brick paving is not recommended in areas subject to freezing. Mortar mix will shrink as it dries, leaving cracks that can collect water. When the water freezes, damage to the paving can result.

1. Thoroughly mix 1 part cement with 3 parts clean sand.

2. Sprinkle mixture over surface. Sweep mixture into the gaps between the bricks.

In next step, do not flood the surface or mortar will stain the surface.

3. Thoroughly wet the surface using a fine spray. Allow moisture to penetrate as deeply as possible without flooding the surface.

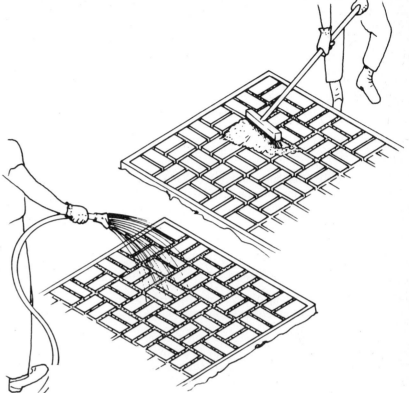

Using Sand

1. Sprinkle dry sand on the surface. Sweep the sand into the gaps between the bricks.

2. Wet the surface to compact the sand.

This procedure may need to be repeated occasionally to keep the gaps filled with sand.

MAKING BRICK WALLS

To ensure that your brick wall will be sound and durable, observe the following rules:

- Be sure that the area is properly graded and drained.

- Make proper bases for footings and slabs.

- Do not mix or work with mortar in cold or freezing weather.

- Be sure that bricks are properly wetted before applying mortar. The damp surface assists the spreading and bonding of the mortar. Because wet bricks do not absorb moisture out of the mortar, the mortar can cure properly. Bricks should be soaked for at least 1 minute before mortar is applied.

The first step in making brick walls is digging and grading. Then make bases, footings or concrete slabs, as required.

If reinforcements are required, they must be carefully spaced and properly embedded in the footing or slab.

Refer to the POURED CONCRETE section for specific information about:

- Grading and building bases and footings.

- Making concrete forms.

Go to next page for instructions for laying bricks.

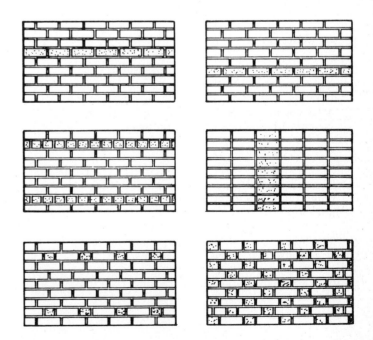

MAKING BRICK WALLS

▶ **Laying Bricks**

Begin laying bricks by making a trial layout of the first course.

1. Establish the exact location of the bricks at each end of the wall.

2. Stretch a mason's line to mark the top outer edge of the first course of bricks. Place line about 1/8-inch away from the surface of the wall to avoid hitting it when laying bricks.

Be sure string is tight and level. If wall is to be a long one, the line may need to be supported at several places to assure correct alignment and height.

3. Place the first course of bricks between the corner bricks. Allow an average of 3/8-inch between the bricks for a mortar joint.

When making the layout, you must take into account the design of the entire wall.

- If the wall is a straight, double-thickness wall the pattern at the end of the wall can be very simple. It can be a column of headers. Headers are bricks laid so that ends [1] rather than sides [2] face the wall.

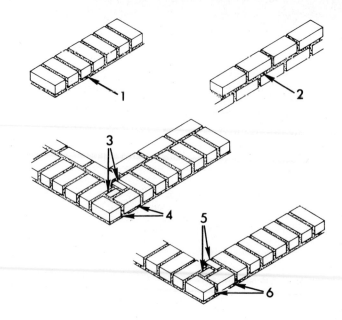

- If the wall is a double-thickness wall that turns a corner, the corner is made of two 1/4 bricks [5] and two 3/4 bricks [6]. The rest of the first course is made of headers.

- If wall is a triple-thickness wall that turns a corner, the corner is made of two 1/4 bricks [3] and two 3/4 bricks [4].

Laying Bricks

If necessary, bricks may be cut, using a bricklayer's set or a mason's chisel. To cut a brick:

- Score the brick on both sides.

- Place the brick on a bed of smooth, wet sand.

- Strike brick at the score mark using a hammer and bricklayer's set.

After you are satisfied with the exact position of each brick in the first course, continue.

4. Remove all bricks.

In next step, make mortar bed as wide as the wall is thick, and long enough to lay several bricks.

5. Spread a bed of mortar [1] approximately 1 inch thick on the foundation.

6. Make edges of mortar bed thicker than the center of the bed.

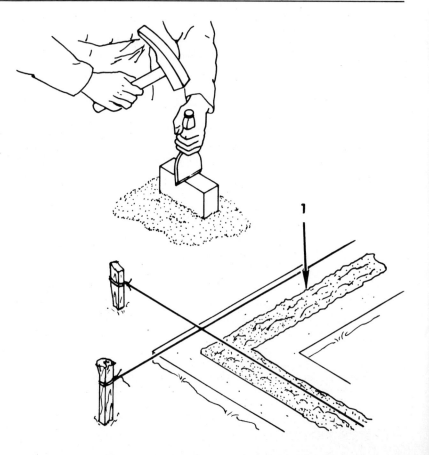

Laying Bricks

7. Lay the corner brick into place in the mortar bed.

If building the corner of a double-thickness wall, lay the bricks as indicated on Page 18.

8. Spread a layer of mortar about 3/4-inch thick on the side or end of the brick next to the corner block. Lay brick in mortar bed next to the previous brick.

9. Repeat Step 8 until several bricks have been laid in mortar bed.

In next step, be sure that all bricks are:

●   Properly aligned with the string and each other.

●   Horizontally level.

●   Vertically plumb.

●   All vertical mortar joints are about 3/8- to 1/2-inch wide.

10. Press all bricks about 1/2-inch into mortar bed.

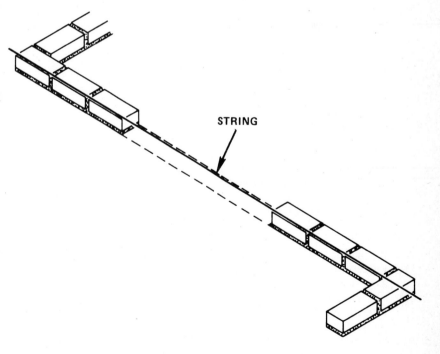

Laying Bricks

11. Cut and remove excess mortar from the mortar joints.

Excess mortar may be spread on the edge or side of the next brick, or be remixed with the mortar on the mortar board.

Repeat Steps 5 through 11 to start the other end of the wall.

After both ends of the wall are laid, the rest of the bricks of the first course may be laid. Work from both ends toward the middle.

When laying bricks in the course, frequent and accurate checks are required to ensure that bricks are:

●   Properly aligned with the string and each other.

●   Horizontally level.

●   Vertically plumb.

●   Mortar joints are not too thick.

STRING

MAKING BRICK WALLS

Laying Bricks

The last brick in a course is called the closure brick. The closure brick must be laid very carefully.

1. Spread a layer of mortar about 3/4-inch thick on the sides [1] or ends [2] of the bricks on each side of the closure.

2. Spread a layer of mortar about 3/4-inch thick on both sides or ends of the closure brick.

3. Carefully slide closure brick into place without knocking the mortar out of the joints.

4. Align closure brick with string and other bricks.

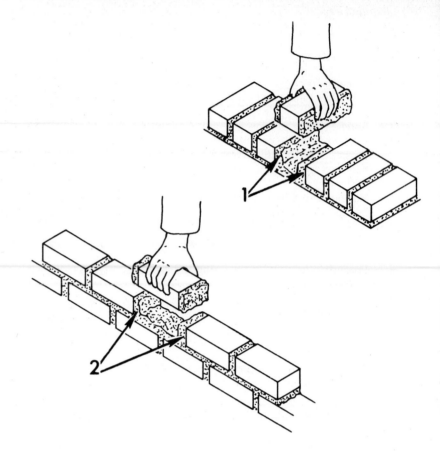

Laying Bricks

After the first course has been laid, the corners or ends of the wall are built. Ends or corners may be built several courses ahead of the stretcher courses. Stretchers are bricks laid so that sides [2] rather than ends [1] face the wall.

When building your brick wall, frequent and accurate checks are required to ensure that:

- Bricks are aligned with the string and each other.

- Courses are horizontally level.

- Wall is vertically plumb.

- The height of each course is checked with a story pole.

If building the corner of a double-thickness wall, the second course [3] and third course [4] is laid as shown.

Remaining stretcher courses follow the patterns shown for the second and third courses.

If additional header courses [1] are required, the pattern and sequence of the first course are repeated.

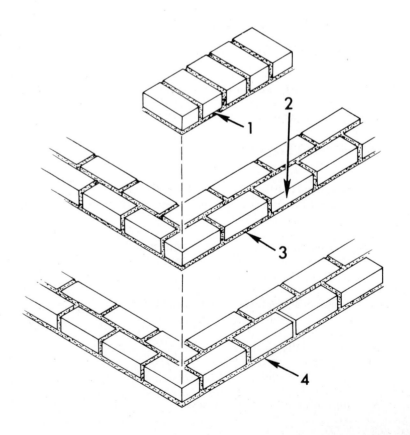

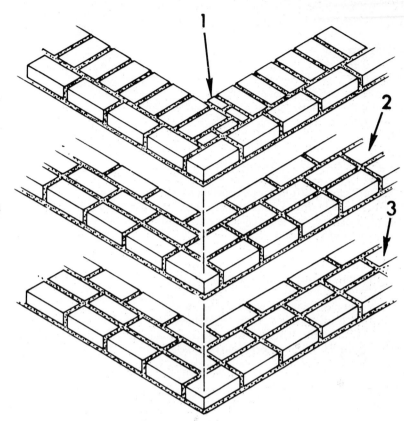

Laying Bricks

If building the corner of a triple-thickness wall, the second course corner is made of four 1/4 bricks [1]. The third course [2] and fourth course [3] are made using whole bricks.

Remaining stretcher courses follow the patterns shown for the third and fourth courses.

If additional header courses are required, the pattern of the first course is repeated.
See Page 18.

▶ **Finishing Mortar Joints**

Mortar joints must be finished after the mortar stiffens but before it hardens.

Several ways of finishing mortar joints are:

- Mortar is troweled flush with face of wall [1].

- A mason's jointer is used to compact the mortar into the mortar joints [2].

- Trowel is used to strike the mortar joint [3]. Struck joints give an even appearance because the top edges of the bricks are carefully leveled and aligned.

- Trowel is used to weather the mortar joint [4]. Weathered joints give an uneven appearance.

- Raked mortar joints [5] are joints from which the mortar has been removed to a desired depth.

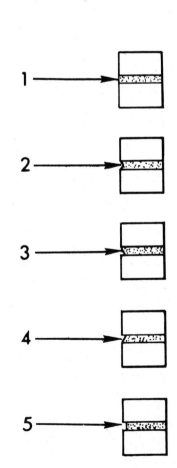

▶ Cleanup

The best way to end up with a clean brick wall is to reduce the requirements for cleanup. Keep the bricks clean of mortar droppings while you are working.

To prevent mortar droppings on the wall, work carefully and keep work area clean.

Do not leave scaffolds in place when they are not in use. Dropped mortar may splash onto the wall. Rain or sprinklers may splatter from the scaffold onto the wall.

If mortar drops on a wall, remove large chunks with a putty knife or wire brush.

After the mortar in the wall hardens, the wall may be cleaned using a commercial masonry cleaning solution. Usually these solutions contain muriatic acid (hydrochloric acid). Follow directions on the label.

The most common problem of soiling on a new brick wall is efflorescence. Efflorescence is a deposit of salts that collect on the surface of brickwork. Moisture passing through the masonry brings the salts to the surface. Usually these salts are white.

Efflorescence can usually be brushed off. More stubborn cases can be removed with a commercial cleaning solution available from your masonry dealer.

Sometimes green stains appear on buff or gray brick. These stains can be removed with soap and water or a mild caustic solution. Do not use an acid solution to clean green stains. Acid will simply change their color to brown.

BRICKWORK REPAIRS

The most common repair to brick work is called tuck-pointing. Tuck-pointing is the repair of masonry joints.

Repairs consist of replacing mortar which has cracked or crumbled.

Generally it pays to tuck-point all the mortar joints at one time. This will repair obviously bad joints and bad joints that are not readily obvious.

Mortar may be removed using a hammer and chisel, but the use of an electric grinding wheel will save time and do a better job.

If the wall is badly damaged, bricks should be removed. The surfaces should then be cleaned and dampened. The bricks are re-installed or new bricks are installed.

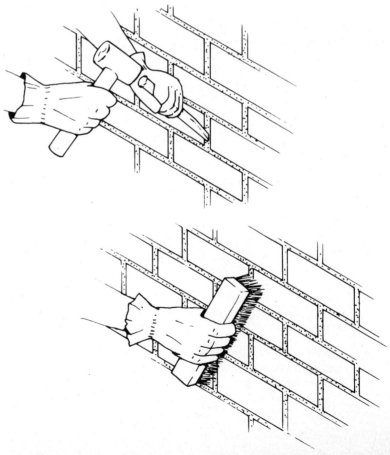

1. Remove existing mortar to a depth of 1/2-inch or until sound mortar is reached.

2. Clean the joints using a brush and clean flowing water.

3.   Dampen the joints.

Tuck-pointing mortar is specially prepared.
Refer to next section (below) for mixing
tuck-pointing mortar.

4.   Using tuck-pointing mortar and a
tuck-pointing trowel [1], fill the joints with
mortar.

The hawk [2] should be held close to the
joint to prevent mortar from dropping onto the
wall.

Tuck-pointing mortar is drier than regular
mortar. For this reason, it is recommended
that the mortar joints be finished while they are
being made.

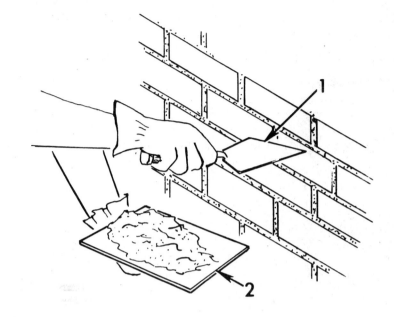

▶ Mixing Tuck-Pointing Mortar

Tuck-pointing mortar is specially prepared to
reduce the effects of shrinkage. This mortar mix
is drier than normal mortar.

For normal use, mix 1 part masonry cement
with 2-1/4 to 3 parts clean sand.

For severe use, mix 1 part masonry cement with
1 part Portland cement and 4-1/2 to 6 parts
sand.

1.   Mix the dry materials thoroughly.

2.   Mix in 1/2 the required amount of water.

Wait one full hour before continuing.

3.   Mix in the rest of the water.

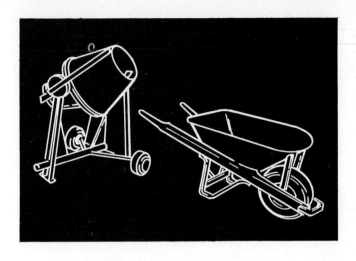

POURED CONCRETE WORK

▶ Description and Uses

Poured concrete is one of the most versatile building materials in use today. It can be made into any shape, with a variety of surface textures and finishes. Colored concrete is also available.

Poured concrete is strong and durable. Because of these properties, it is an excellent material for building permanent structures around your home. Driveways, sidewalks, patios and steps are just a few of the projects which you can do with concrete.

However, when considering a concrete addition, do not take the job lightly. A considerable investment of time and money may be required for your project to achieve professional results.

It is probably a good idea to check with a landscape designer and a concrete contractor before beginning. They can help you determine specific design and planning needs.

A landscape designer can give you ideas for making the design of your concrete compatible with your home and other existing structures.

The shape and surface texture of your concrete can be varied to either match or contrast with features of your home. Dividers, borders and edgings of wood or other materials may be used to provide accent to the concrete.

A concrete contractor can give you tips on specific construction problems. You may even want to hire him to do the job. For example, it is recommended that you hire a contractor if any of the following soil conditions exist:

- A ground water condition.

- Unstable or expansive soil.

- Concrete is to be poured over 3 feet of earth fill or over 6 feet of sand or gravel fill.

A concrete project is likely to be covered by requirements of local building codes, the Federal Housing Authority (FHA) and the Veteran's Administration (VA). The next section (below) describes the requirements and guidelines which are usually provided by these codes and agencies.

▶ Building Code, FHA and VA Requirements

The requirements of building codes, the FHA and the VA apply to the design and construction of many home concrete projects. Specifications published by government agencies are excellent sources of specific directions.

Various local agencies may be responsible for building codes and regulations. They also issue building permits when required and furnish inspectors to ensure that the work conforms to requirements.

Building codes and specifications are likely to provide specific direction on the following types of subjects:

- Standard slopes for concrete surfaces
- Preparation of the subgrade
- Specifications for building forms
- Concrete composition
- Concrete thickness
- Coloring restrictions
- Requirements for expansion joints and control joints

- Requirements for reinforcement
- Concrete curing and hardening requirements

From the preceding list, it is apparent that building codes and specifications can be a big help in designing and constructing your project.

In the design and planning of your concrete addition, it is also essential to consider both FHA and VA minimum property standards or construction requirements.

Contact the local office of these agencies for published standards. In order to guarantee future mortgages on your property by the FHA or VA, you must make your concrete addition meet their requirements.

Poured concrete work requires considerable planning to obtain professional results. Basic to any planning is consideration of weather conditions for your work.

Plan your work for a cool, but not cold, time of the year. Concrete must not be allowed to freeze before it hardens. Nor should concrete be poured on frozen ground.

In hot weather, concrete will quickly dry out, requiring you to work very rapidly after pouring it.

Another basic planning consideration is to account for future architectural and landscaping changes. Because of the permanency of concrete, you must provide for the following types of changes before beginning your concrete work:

- Extend water lines for hose outlets, sprinkler systems, plumbing lines, etc.

- Extend electrical wiring.

- Extend gas lines.

To effectively plan your job, you should become familiar with the operations required for a typical concrete project. The operations are described in the following steps and arranged in the sequence of performance.

1. Dig and grade the subbase. Page 43.

2. Build a base, usually of crushed rock, if required by soil and drainage conditions. Page 44.

3. Build forms. Page 45.

4. Prepare reinforcements if required. Page 49.

5. Estimate the amount of concrete needed. Page 51. This can also be done before Step 1. Be sure to read Pages 28 through 31, Types of Concrete and Concrete Finishes, before buying your concrete. Materials vary depending on the type of concrete and finish you choose.

6. Buy your concrete. Page 52 describes the different ways concrete can be purchased.

7. Get things ready for mixing and placing concrete. Page 54.

8. Mix concrete if you are doing it yourself. Page 55.

9. Place concrete. Page 58. Pouring, spading, striking off, tamping, bullfloating and darbying are included in this operation.

10. Finish concrete. Page 61. Your desired surface finish is made at this time. Edging, making control joints, floating and troweling may be included in this operation, depending on the type of project and desired finish.

11. Allow concrete to cure. Page 64. Curing keeps the concrete moist for a period of days to ensure quality, hardness and durability.

After becoming familiar with the operations involved with concrete work, read the recommended design considerations for the specific project you have in mind. These project descriptions begin on Page 32. Be sure to check your local building codes to plan your job within their applicable requirements.

The best method of completing your plans is to make a scale drawing of your project. A piece of graph paper is handy for making the plans.

Include on your drawing the dimensions of your project, slope requirements, concrete thickness and thickness of your base, if required. Add any other information necessary to aid you in completing your project.

The drawing is used as a ready reference for estimating the amount of supplies and concrete materials you need. It includes specifications for digging, grading, sloping and building forms.

TYPES OF CONCRETE

Concrete is a mixture of Portland cement, sand, coarse aggregate (gravel, pebbles, etc.) and water. The proportions of these ingredients vary depending on the requirements of your project.

There are three types of concrete recommended for your use:

- Plain concrete.

- Early-strength concrete.

- Air-entraining concrete.

Of the three, air-entraining concrete is generally recommended. It contains an additive which forms tiny air bubbles in the hardened concrete. Consequently, the chances of cracking due to freezing and thawing are reduced.

Plain concrete forms a slightly harder surface than air-entraining concrete. However, it is somewhat more difficult to make a smooth finished surface. Plain concrete is recommended for small concrete jobs, providing you live in a moderate to mild climate.

Early-strength concrete hardens quicker than the other types of concrete, thus requiring less time for the curing process. However, it is generally not as strong as plain concrete. Air-entraining additives can be added to early-strength concrete if desired.

CONCRETE FINISHES

A wide variety of concrete finishes is available. Attractive contrasts can be made by using any combination of finishes.

The type of finish and the method of achieving it must be determined in the planning stage of your work. Mixing requirements and finishing techniques vary depending on the type of finish you choose.

The following concrete finishes and the methods for achieving them are included in this section:

Plain finishes [1]. Page 29.
Rough finishes [2]. Page 29.
Decorative marks [3]. Page 29.
Pebble finishes [4]. Page 30.
Colored finishes. Page 31.

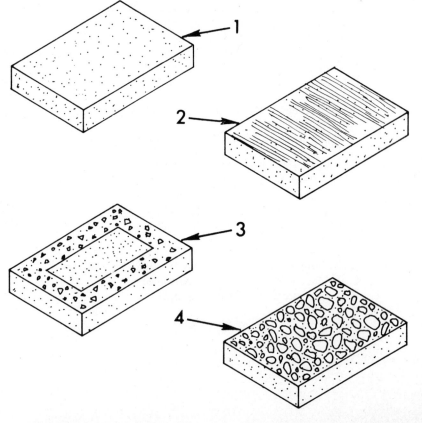

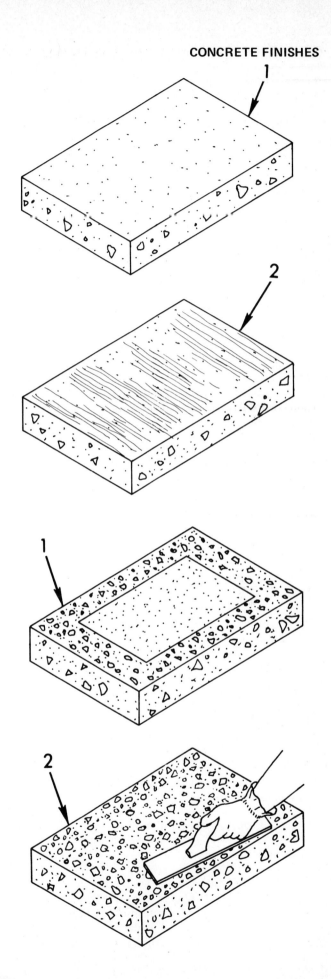

▶ **Making Plain Finishes**

Plain, smooth finishes [1] are made as follows:

1. Place concrete in the normal manner.

2. Finish concrete with a steel trowel to produce a smooth surface. Allow concrete to cure.

▶ **Making Rough Finishes**

Rough finishes [2] provide a non-slip surface. The roughness depends on the type of tool you use during the finishing operation. Brushes and brooms of varied stiffness can be used.

Rough finishes can be made in any pattern or direction. However, for sidewalks and driveways, brush strokes are normally made perpendicular to the flow of traffic.

1. Place concrete in the normal manner.

2. Instead of using a steel trowel to finish the concrete, use a brush or broom to make your rough finish. Allow concrete to cure.

▶ **Making Decorative Markings**

Decorative markings [1] such as leaf impressions, circles, edgings and rock salt indentions can add a unique touch to your concrete.

Decorative markings are normally made after finishing the concrete with a steel trowel to produce a plain finish.

<u>CAUTION</u>

Do not indent concrete too deeply in areas subject to freezing. Water can freeze in deep markings, causing possible cracking and flaking.

Impressions and edgings can be made with whatever tools and materials you like. Make rock salt finishes [2] as follows:

1. After troweling the surface the first time, scatter rock salt as required.

2. Using the steel trowel, lightly press rock salt into surface.

3. When concrete has completely cured, wash rock salt from surface.

▶ Making Pebble Finishes

Pebble finishes are some of the most attractive concrete finishes made. Many pebbles or just a few can be used to create different effects.

There is a wide variety of pebble sizes and colors from which to choose. However, when you buy pebbles, be sure to get rounded stones, not chips or slivers.

There are two common methods of making pebble finishes:

> Using pebbles [1] in place of coarse aggregate when mixing concrete. See section below.
>
> Placing pebbles [2] on the surface of a plain concrete finish. Page 31.

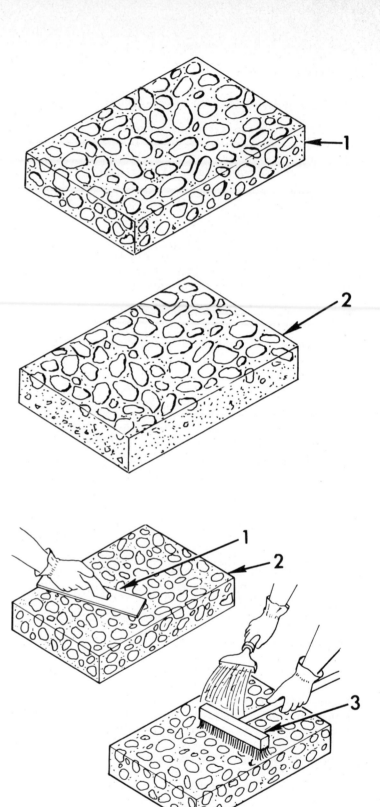

Making Pebble Finishes

Mixing Pebbles into Concrete

1. Using pebbles instead of coarse aggregate, mix concrete. Use less sand to make the consistency of the concrete mixture more firm.

A firm mixture is necessary to prevent the pebbles from settling when the concrete is poured.

2. Place the concrete.

CAUTION

When using float [1], do not work the pebbles [2] too far below the surface.

3. Using float [1], finish the surface. Do not use a trowel. Allow surface to begin to harden.

4. Using a stiff brush [3] and hose, remove the top third layer of concrete from the pebbles. If pebbles are brushed off surface, allow concrete to become harder.

5. Repeat Step 4 until top third layer of cement is removed from entire surface and water runs clear.

6. Allow concrete to cure.

Making Pebble Finishes

Placing Pebbles into Plain Finishes

1. Mix concrete using coarse aggregate as
    normal. Use less sand to make the
    consistency of the concrete mixture
    more firm.

A firm mixture is necessary to prevent pebbles
from settling into the concrete when they
are placed.

2. Pour concrete slightly below tops of forms
    to allow for addition of pebbles.

If using large pebbles, place them flat side up.

3. After the darbying operation, Page 60,
    place pebbles [2] onto the surface. Using
    a wood float [1], gently press them just
    below the surface of the concrete.
    Allow surface water to disappear.

CAUTION

**When using float, do not work the pebbles too
far below the surface.**

4. Using float [1], finish the surface. Do not
    use a trowel. Allow surface to begin to
    harden.

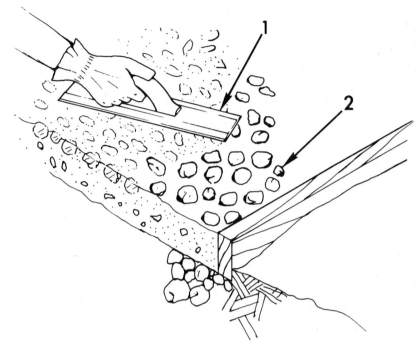

5. Perform Steps 4 through 6 on Page 30 to
    expose and wash the surface.

▶ **Making Colored Finishes**

Several options are available for making colored
concrete. One method is to use paint or stains
after the concrete is cured completely. The
paint or stain must be the type made for
concrete surfaces.

More permanent methods involve adding color
pigments when the concrete is mixed.

CAUTION

**Be sure to follow manufacturer's instructions
for using color pigments. Too much coloring
can weaken your concrete.**

There are two methods of adding color pigment.
Because the coloring is fairly expensive, the
second method which uses less pigment may
be preferred.

Either method requires mixing concrete in a
cement mixer. Mixing by hand cannot produce an adequately uniform color. Be sure
the cement mixer is clean before mixing.

Coloring the Entire Concrete Thickness

1. Mix color pigment with white Portland
    cement in the dry state. If a black or grey
    color is desired, the cement does not have
    to be white. Use a light-colored sand.

2. Mix concrete with pigment added.

3. Place, finish and cure concrete as normal.

Coloring a Thin Top Layer of Concrete

1. Mix and pour uncolored concrete to within
    1/2-inch of the tops of the forms. Spade
    and tamp the concrete.

Do not allow uncolored layer to begin to harden
before pouring colored layer.

2. Mix a colored mortar consisting of the
    color pigment, 1 part white Portland
    cement and 1-1/2 parts light-colored
    sand. No coarse aggregate is required.

3. Place the colored mortar over the layer of
    concrete. No spading and tamping is
    required.

4. Finish and cure concrete as normal.

▶ Driveways

The design of a driveway depends largely on your own personal requirements. Whether you plan a one-or-two-car driveway, space for guest parking [1], or a circular drive-through driveway [3], you should account for as many present and future needs as possible.

Building codes are likely to contain design and construction information, especially for the part [2] between the street and sidewalk.

Be sure to check building codes for general guidance and specific requirements. A building permit and inspections may be required.

The following design considerations are recommended for your driveway:

Thickness [4]: Concrete thickness for passenger cars only should be 4 inches minimum. A 5- or 6-inch thickness is recommended for heavier vehicles. Thick driveways provide long-term durability.

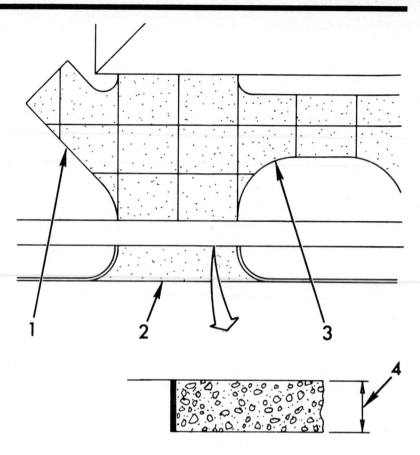

Driveways

Width [1]: Allow a 10-foot width for each car. An ideal two-car driveway should be 20 to 24 feet wide. Make curved sections [3] wider than straight sections. The street entrance [2] is usually flared to provide safe entrance and exit.

Slope [4]: The driveway should slope away from the house for proper water runoff. It should allow clearance for your car at the street entrance, with no humps along the surface to scrape the bottom of your car. Generally, the slope should be no more than 1-3/4 inches per foot. A minimum slope of 1/4-inch per foot is recommended. A cross-slope [5] across the width of the driveway should be built from 1/8- to 5/8-inch per foot.

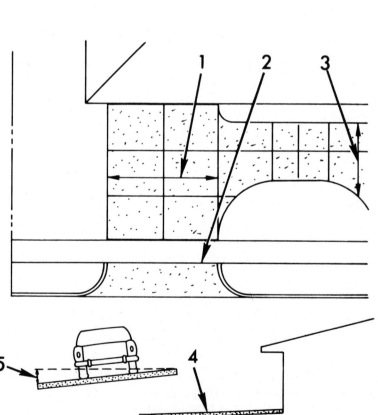

Driveways

Expansion joints [1]: Make expansion joints where the driveway meets existing structures such as garage floor, sidewalk and curbs.

Reinforcement: Reinforcing is optional. However, building codes may provide guidelines for its use.

Control joints [2]: Make control joints at 10-foot intervals maximum. Ideally, they should be made to separate the driveway into 4-to-6-foot square panels.

Concrete finish: Building codes may specify the type of finish. The finish should be rough, with the texture perpendicular to traffic flow.

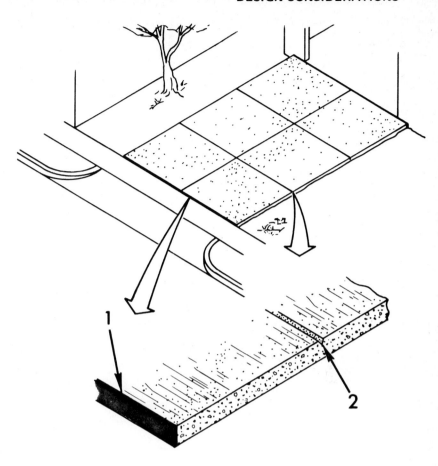

▶ **Sidewalks**

The design of a sidewalk will vary depending on whether you plan a public sidewalk [1] or a service or access sidewalk [3] for your own use only.

Building codes are likely to contain design and construction information for public sidewalks. These requirements may also serve as a guide for building service or access walks.

Check your local codes for sidewalk specifications. A building permit and inspections may be required.

The following design considerations are recommended for your sidewalk:

Thickness [2]: Concrete thickness should be 4 inches minimum.

Width: The minimum width of a public sidewalk [1] or one leading to the entrance of your home should be 4 feet. It should be wide enough to allow three people to walk abreast. Service or access walks [3] are a minimum of 2 to 3 feet wide.

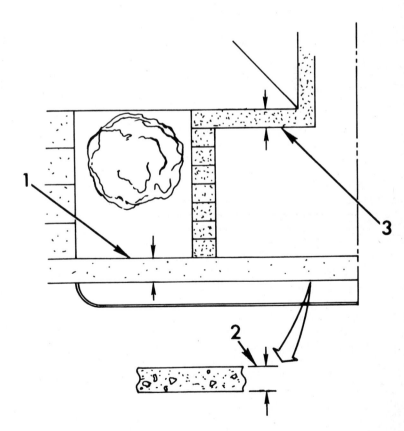

DESIGN CONSIDERATIONS

Sidewalks

Slope: A sidewalk should be sloped to allow proper water runoff, yet not so steep as to be a hazard. In areas subject to freezing, the maximum slope should be 5/8-inch per foot. In areas not subject to freezing, allow a maximum slope of 1-3/4 inches per foot. A cross-slope [3] across the width of the sidewalk can be built from 1/8- to 5/8-inch per foot.

Expansion joints [1]: Make expansion joints where the sidewalk meets existing structures such as home entrances, driveways and patios.

Reinforcement: Reinforcing sidewalks is optional. However, building codes may provide guidelines for its use. Generally, reinforcement is not required with properly made control joints.

Control joints [2]: Make control joints at intervals equal to the width of the sidewalk. A maximum of 4-foot intervals is recommended.

Concrete finish: Building codes may specify the type of finish. The finish should be rough, with the texture perpendicular to traffic flow.

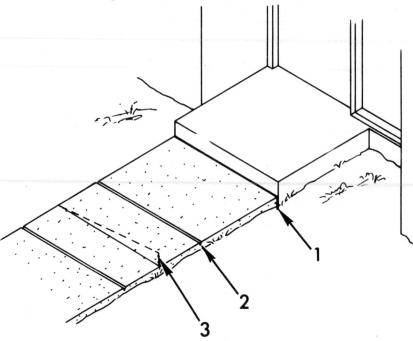

▶ **Patios**

Patios can be built to your own specific design preferences. You can plan for a variety of shapes, sizes and surface finishes.

Building codes are not likely to apply, although they may be useful as a reference for general concrete requirements in your area.

The following design considerations are recommended for your patio:

Thickness [1]: Concrete thickness should be 4 inches minimum.

Slope: The patio should slope away from your home for proper water runoff. A minimum slope of 1/8-inch per foot is recommended.

Forms: An alternate to standard removable forms is the use of redwood forms [2] built as a permanent part of your patio. Any variation of the forms shown can be used. You may want to treat the wood with a varnish or wood preservative before placing concrete. For protection during placing and finishing concrete, masking tape can be applied along the top edges of the forms.

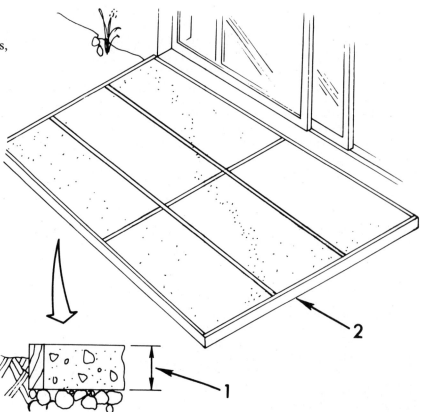

34

Patios

Expansion joints [1]: If using standard, removable forms, make expansion joints where the patio meets existing structures. Build joints around any plumbing and utility fixtures.

Anchor bolt provisions: If planning to add a patio cover, you may want to install anchor bolts for attaching the patio cover supports. Install bolts before finishing concrete.

Reinforcement: Reinforcing is optional. However, building codes may provide guidelines for its use.

Control joints [2]: If using standard, removable forms, make control joints at 10-foot intervals maximum. Ideally, they should be made to separate the patio into 4-to-6 foot square panels.

Concrete finish: Any textured finish is recommended. A non-slip surface should be provided.

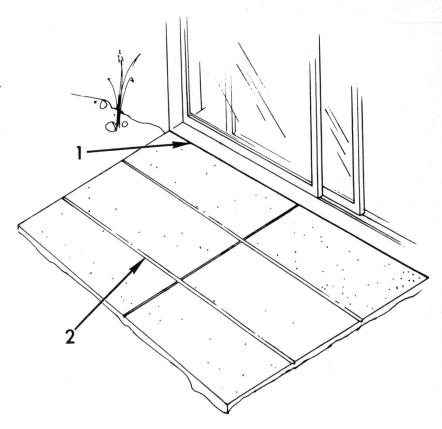

▶ **Foundation Slabs**

Foundation slabs can be built as the substructure for house additions, garages, storage sheds, outdoor barbeques and many other building projects. Depending on the purpose of the slab, building codes and ordinances are likely to specify required building permits and inspections.

Be sure to check your codes for general guidance and specific requirements for your project.

The following design considerations are recommended for your foundation slab:

Thickness [1]: Concrete thickness should be 4 inches minimum.

Slope: The slope of your slab depends on its intended use. No slope is required if provisions for water runoff are not needed. The slab must be level if building structures on it. If a slope is required, it should be a minimum of 1/8-inch per foot.

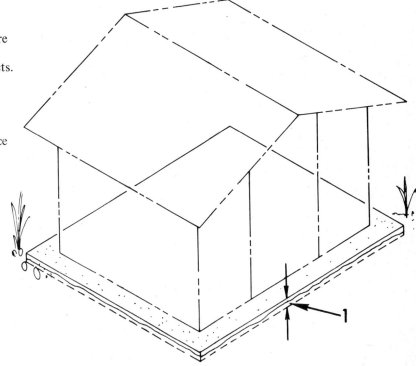

Foundation Slabs

Expansion joints [1]: Build expansion joints where the slab meets existing structures. Build joints around any plumbing and utility fixtures.

Anchor bolt provisions [2]: You may want to install anchor bolts in the slab, depending on the type of structure you plan to build. Install bolts before finishing concrete.

Reinforcements: Reinforcement requirements depend on the use of the slab. Building codes may provide guidance. For large slabs, welding wire fabric [3] is recommended. If masonry walls are to be built, steel reinforcing rods [4] should be installed in the slab at appropriate intervals when concrete is poured.

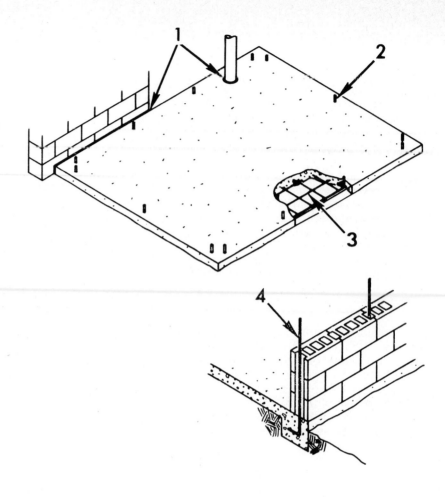

Foundation Slabs

Control joints [1]: If welding wire fabric reinforcement is used in the slab, control joints may not be required. However, if you make control joints, they should be at 10-foot intervals maximum. Ideally, they should be made to separate the slab into 4-to-6 foot square panels.

Concrete finish: Depending on the use of the slab, the concrete finish may vary. For slabs which will be covered with floor material, a plain smooth finish is recommended. Any non-slip surface should be made for slabs which will carry traffic.

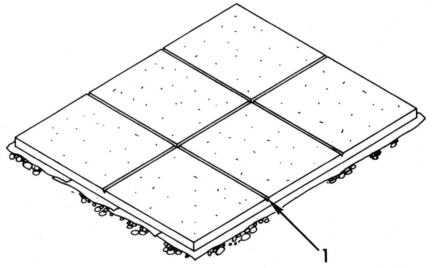

▶ Footings

Concrete footings [3] are required below walls, steps and any other structures which need protection from settling into the ground. Unless you build on a foundation slab, footings are mandatory for this purpose.

Building codes are an excellent source of design and construction specifications. They also describe what footings need building permits and inspections. Be sure to check codes for general guidance and specific footing requirements.

Footing design depends on the structure being built. You must plan footing and structure requirements at the same time.

For example, suppose you plan to build a concrete block wall [1] supported by a footing. The alignment of the footing, its width and length, and spacing of reinforcement rods [2], if required, must be determined when you plan the wall.

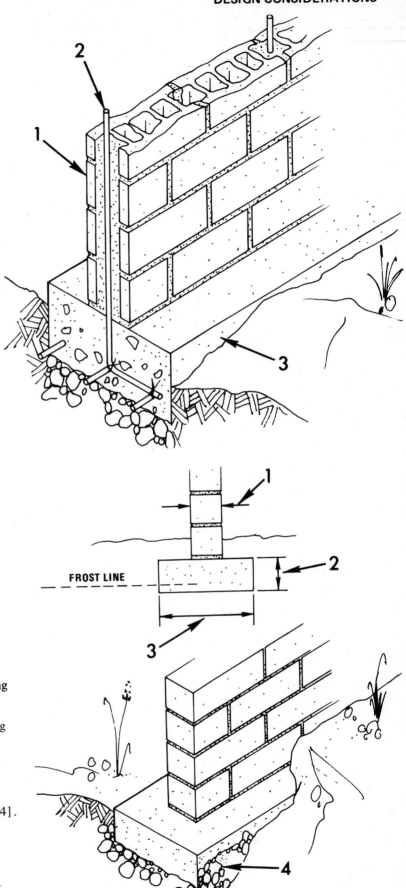

Footings

The following design considerations are recommended for footings:

Thickness: Concrete thickness [2] is normally the same as the width [1] of the structure being supported.

Width [3]: For walls, a footing should be twice as wide as the wall is thick. For steps, a 6-to-8-inch width is usually adequate.

The width may vary in the same footing. For example, a retaining wall may require a footing wider at the bottom for proper support.

Digging and Grading: The bottom of the footing must be below the frost line. Depth of frost line depends upon severity of local freezing conditions. Check local building codes for specific depth. The top of the footing must be at or below ground level. Footings normally require a 6-inch base of crushed rock [4]. Digging must account for both the base and the footing.

Grading must be done carefully to provide the correct alignment for the footing and structure.

DESIGN CONSIDERATIONS

Footings

Drainage: Drainage provisions must be made, usually by adding drain pipes [2] along the footing during digging and grading. For long footings, drain pipes may be required through the footing itself.

Forms: Forms must be built so that their tops accurately match the slope requirements of the structure being built. Footings may require no forms if the soil is firm. Make the form when you dig and grade. Use the soil edges as your forms.

If you require a footing with forms more than two or three boards high, the design considerations for wall forms apply. Page 42.

Reinforcement: A minimum of three steel reinforcing rods [1] should be placed horizontally in a footing. They should be placed one-third of the way up from the bottom.

Vertical reinforcing rods are required for most walls. They should be placed into the concrete at appropriate intervals for the type of wall you plan.

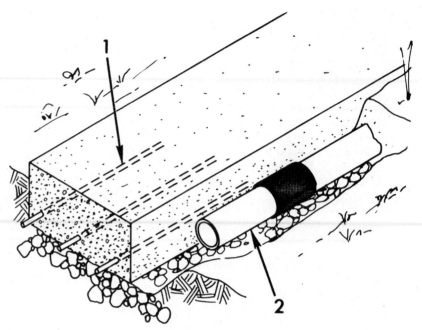

Footings

Reinforcement: In addition to the reinforcing rod requirements, footings for poured concrete walls should be built with a keyway [1] to help secure the wall to the footing. Page 39 describes how to make the keyway and install vertical reinforcing rods in a footing.

Placing concrete: Pour concrete for footings in even layers of no more than 24 inches thick. Spade and tamp each layer before pouring next layer. Be sure reinforcing rods [2] remain in place. Do not allow one layer to begin to harden before pouring the next.

Curing: Allow concrete to cure completely before building your structure.

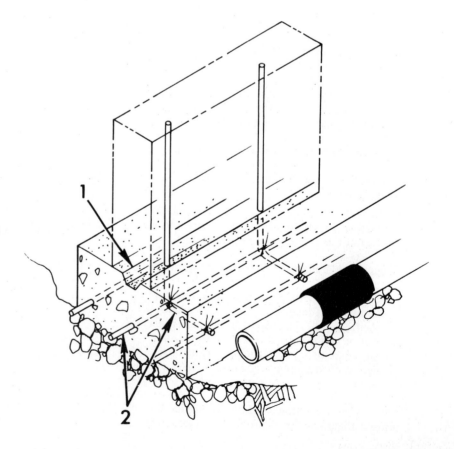

▶ **Making a Footing Keyway**

A keyway [6] is made to provide an anchor for
a poured concrete wall. Depending on reinforcement requirements, vertical reinforcing rods [1]
may also be installed when you make the keyway.

1. Obtain a length of 2 x 4 lumber [2] equal
    to the length of the footing. Taper the
    2 x 4 as shown.

2. If installing reinforcing rods [1], drill
    holes [4] in the center of the 2 x 4 at
    intervals equal to the spacing requirements
    of the rods. Holes should be the same
    diameter as the rods. For poured concrete
    walls, unless otherwise specified by code,
    drill holes 24-inches apart.

3. Saw the 2 x 4 in half, along the centerline [5]
    of the holes [4].

4. Pour concrete to 1 inch below the top of
    the footing forms [3]. Place 2 x 4 halves [2]
    into the concrete, making sure they are
    properly aligned and positioned for the
    structure.

5. Pour remaining concrete to the tops of the
    forms [3].

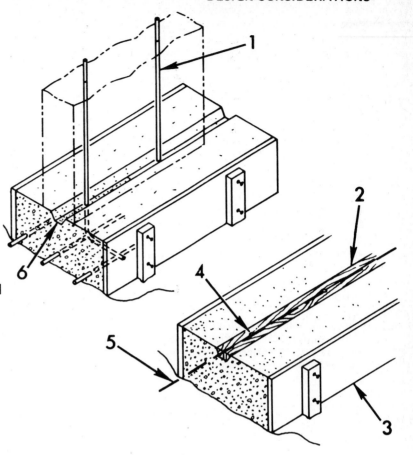

Making a Footing Keyway

6. Immediately after pouring concrete, care-
    fully push reinforcing rods [1] through
    holes [2] in the 2 x 4. Be careful not to
    disturb the position of the horizontal
    rods [4].

7. Spade and tamp the concrete.

8. After concrete is hard enough to support the
    vertical rods [1], remove 2 x 4 halves [3].

9. Allow concrete to cure.

Immediately before pouring the concrete wall,
mix a thin paste of cement and water. Apply the
paste over the entire surface of the keyway [5] to
provide a good bond for the concrete wall. Do not
allow the paste to begin to dry before pouring the
first layer of the wall.

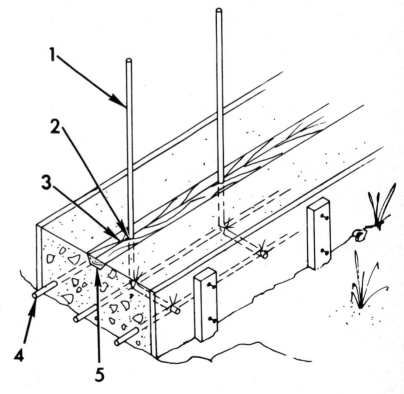

DESIGN CONSIDERATIONS

▶ **Steps**

Steps can be built at entryways to your home, to change levels in your garden or at any other location you desire. Steps consist of landings [1], risers [2] and treads [3] as required by your design.

It is very likely that building codes will contain specific design information for step construction. Some steps may require building permits and inspections.

The following design considerations are recommended for your steps:

Landings [1]: Make landings level from side to side. Landings should be 1/4-inch higher where they meet the entryway. They should extend well beyond the door sill.

Risers [2]: Make risers of uniform height and level from side to side. Height dimensions are usually code specified depending on the number of steps desired.

Treads [3]: Make treads of uniform width from front to back and level from side to side. Treads should be 1/4-inch higher at the back than at the front edge.

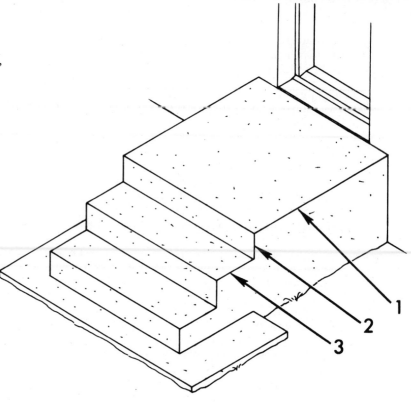

Steps

Footings [3]: Footings are required, designed to match the step structure.

Expansion joints [1]: For steps at an entryway, make a joint, called an isolation joint in this case, between the steps and the building. A layer of building paper is commonly used.

Forms: Two methods can be used for building forms: build forms to remain in place until concrete is cured, or build forms to be removed after concrete hardens enough to support itself.

If forms are removed, vertical concrete surfaces can then be finished to the desired texture. If forms remain in place, vertical surfaces will be the same texture as the forms.

Riser forms [2] which remain in place should be beveled along the bottom edge to allow finishing the tread.

If your step forms are more than two or three boards high, the design considerations for wall forms apply. Page 42.

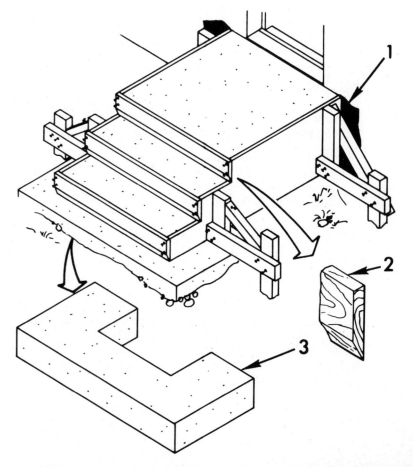

Steps

Anchoring provisions: Steps may require tie
rods [1] for anchoring to an existing structure.
Check building codes for specific requirements.

Reinforcement: Reinforcing steps is optional.
However, building codes may provide guidelines for its use.

Placing concrete: Pour concrete for steps in even
layers of no more than 24 inches thick. Spade
and tamp each layer before pouring the next
layer. Do not allow one layer to harden
before pouring the next.

Concrete finish: Any textured finish is recommended. A non-slip surface on
landings [2] and treads [3] should be
provided.

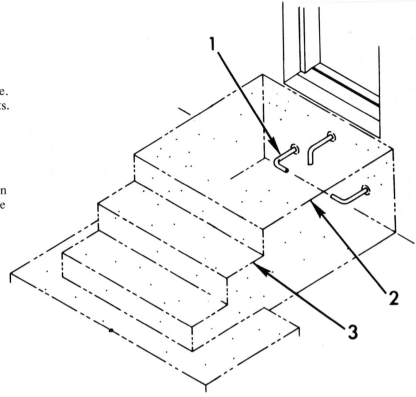

▶ Walls

The design of a poured concrete wall depends
largely on its purpose. Retaining walls, partition walls and foundation walls can all be built
with poured concrete.

However, building codes and ordinances are
likely to specify strict requirements for concrete walls. Before planning a wall, be sure to
check your codes for design and construction
criteria. Building permits and inspections will
probably be required.

The following design considerations are recommended for concrete walls. Use them as general guidance for wall construction.

Footings [2]: Footings are required, designed
to match the wall structure. Be sure to build
the footing level and at the correct alignment
for your wall. A keyway [3] and vertical
reinforcing rods [1] are required. Unless
otherwise specified by code, rods should be
installed at 24 inch intervals. Page 39 describes
how to make the keyway and install vertical
reinforcing rods in a footing.

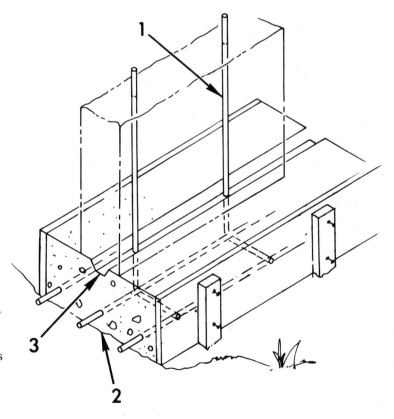

Walls

Forms: Forms for walls must be strong, straight and watertight. One-inch tongue and groove boards [2], plywood or plyscord is recommended. Plyscord is plywood which is smooth on one side for producing a smooth concrete finish.

Forms must be adequately supported and braced, strong enough to support the pressure of the wet concrete until it hardens.

Removable wood spreaders [3] are used to separate the forms until concrete is poured. Cross-ties [1] help hold the forms together. Wire ties [4] can be added for more support.

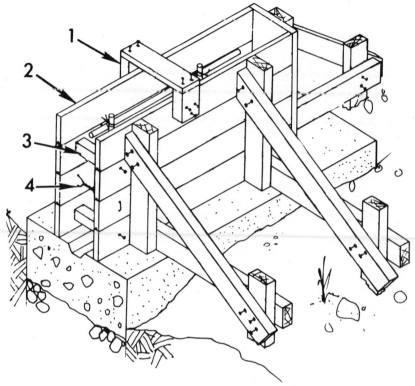

Walls

Reinforcement: In addition to the vertical reinforcing rods [3] extending from the footing, horizontal rods [4] or welding wire fabric [2] are often used. Be sure to check building codes for specific requirements.

Both reinforcements [2, 4] are tied to the vertical rods [3]. Welding wire fabric [2] should also be tied to the forms. Removable wood blocks [1] are used to provide spacing between the wire and the forms.

Placing concrete: Pour concrete for walls in even layers of no more than 24 inches thick. Begin pouring at the ends of the wall. Spade and tamp each layer before pouring the next layer.

Be sure reinforcements [2, 4] remain in place. Do not allow one layer of concrete to begin to harden before pouring the next.

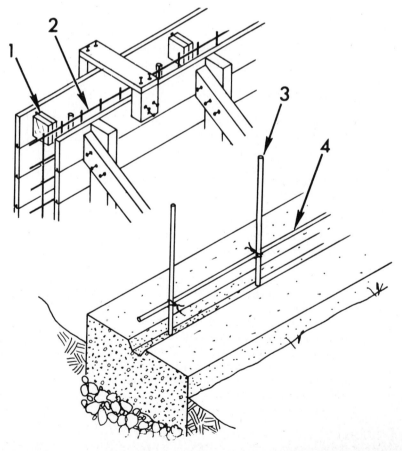

▶ Digging and Grading

Digging and grading is the first step of any concrete project. Your subbase must be at the required depth and slope for concrete work.

Depending on the size of the job, digging and grading can be accomplished by hand or by a tractor with a scoop or other power equipment.

A good rule to follow is to hire a contractor with power equipment if it would take two men working with shovels and wheelbarrows more than one day to remove the dirt. When digging down more than 4 to 6 inches, you rapidly start dealing with large, heavy quantities of soil, especially for big jobs.

If you hire a contractor, lean heavily on his experience to obtain the proper depth, slope and dimension requirements for your project. Place stakes in the ground to mark the outline as an aid for digging and grading.

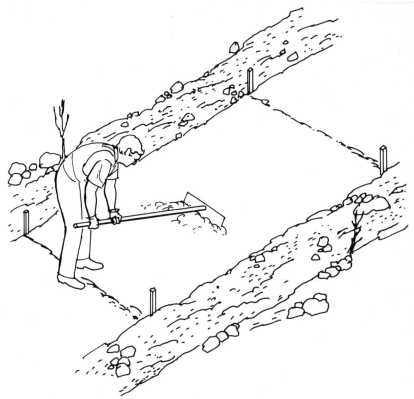

Digging and Grading

In addition to any specific digging and grading requirements described in Design Considerations, Pages 32-42, digging and grading must be accomplished as follows:

- A smooth subbase [2] must be provided.

- Be careful not to disturb the soil unnecessarily. Loose soil must be thoroughly compacted with a dirt tamp [1] or roller to provide a firm subbase.

- The slope of grading must match the required slope of your concrete [4]. A transit level or hand level can be rented to obtain the proper slope. Ask your dealer for operating instructions for the level.

- The depth of grading [3] must allow for the concrete thickness [4] and any required base [5]. Base requirements are described on Page 44.

- If a base [5] is required, it must be built larger than the dimensions of the concrete, requiring extra digging and grading.

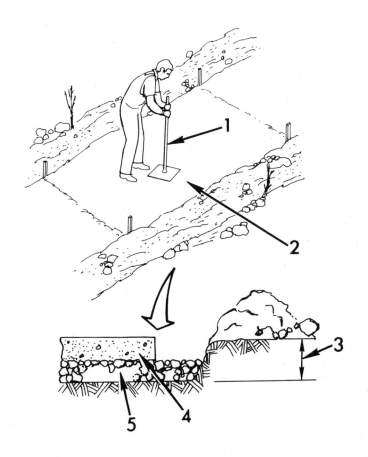

PREPARATION FOR POURED CONCRETE

Digging and Grading

- On clay or poorly drained soil, digging and grading must allow for drain pipes [1, 3] along the base. Four-inch tile pipes are recommended, spaced 1/4-inch apart. Building paper [2] is wrapped around the spaces between pipes.

- If drain pipes [1, 3] are required, a transit level or hand level can be used to obtain the required slope of grading.

- Rocks just at or below the graded level must be removed.

- All roots must be removed.

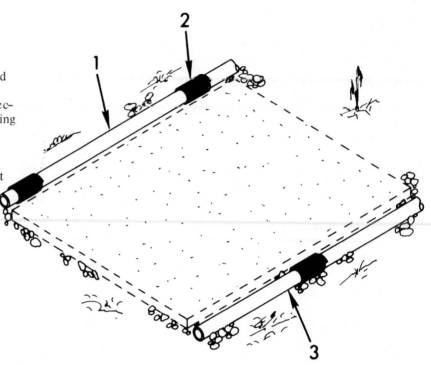

▶ Building Bases

A crushed rock or gravel base [1] is often required under concrete to provide adequate support and water drainage. It is an excellent way of ensuring that your concrete stays in first-rate condition for as long as possible.

The need for a base of crushed rock usually depends on local soil conditions.

On filled soil, whether it is well drained or not, a base is required. The subbase must be thoroughly compacted with a dirt tamp or roller before placing the crushed rock. Thickness of the base should be 6 inches minimum.

On clay or poorly drained soil, a base is required with a minimum 2-inch thickness.

On sandy, well-drained soil, a base is not generally required.

If in doubt about your soil conditions and whether your project requires a base, check with a building inspector. Building regulations usually specify requirements for bases.

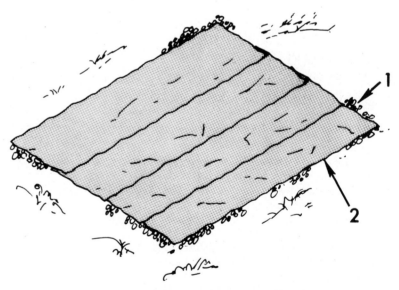

Be sure to allow for the thickness of the base [1] when digging and grading. Page 43. It must be smooth and of equal thickness over the entire grade.

After placing the crushed rock base [1], cover it completely with tarpaper [2] or plastic sheeting. This covering prevents water from leaving the concrete mixture after it is poured.

▶ **Building Forms**

Forms are built as a sturdy mold for pouring
concrete. The tops of the forms are used as a
guide for operations required in placing and
finishing concrete.

Instructions in this section provide general guidelines for building forms. Specific information on
forms for your project is described in Design
Considerations, Page 32, where applicable.

Forms are usually one of a kind, built to meet
the exact thickness, slope and dimension
requirements of your concrete project. They
consist of lengths of lumber [1] supported by
stakes [2] and braces [3] as required.

Lumber for forms should not be dry or fully
seasoned. Dry lumber is likely to absorb
water from the concrete mixture. Improper
curing results.

Fir, pine, spruce and redwood are recommended
woods for building forms. They are all
relatively light and easy to work with.

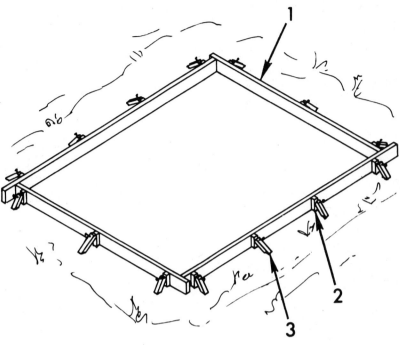

Building Forms

The size of lumber used to build forms depends
on the requirements of your project. 1 x 4's or
2 x 4's are used for 4-inch thick concrete [3].
1 x 6's or 2 x 6's are used for 6-inch thick
concrete [5].

For 5-inch thick concrete [4], 6-inch lumber
is recommended, placed into the subbase to obtain the required 5-inch height. Four-inch lumber
can be used. The space below the form is
packed firmly with soil.

Because lumber used for building forms is
undersize (1 x 4's are actually a fraction
less than 1 inch by 4 inches), be sure the
installed height of the forms meets the requirements for your concrete thickness.

Stakes [1] and braces [2] can be any size
required. They can be purchased precut, or you
can make them yourself. Usually 1 x 2's,
2 x 2's or 2 x 4's are best.

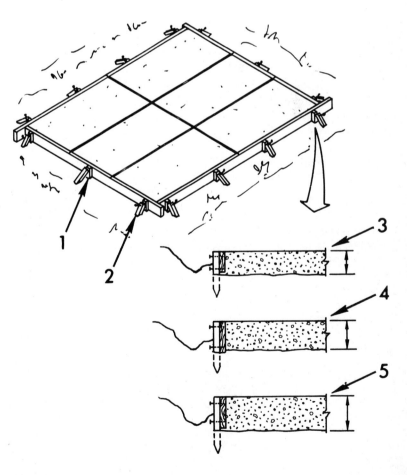

Building Forms

To obtain the proper slope of your forms [1], a transit or hand level can be rented. Ask the equipment rental dealer for operating instructions for the level.

Forms [1] must be built sturdy enough to support the pressure of the poured concrete until it hardens. If using 2-inch lumber, stakes [2] can be placed as far apart as 4 feet.

For 1-inch lumber, place stakes [2] at intervals of no more than 2 feet. Install braces [3] as required. Double-headed nails [4] should be used to make form removal easier.

If forms [1] are not supported adequately, they can lean or collapse when concrete is poured. It is a good idea to have extra stakes [2] and braces [3] handy before pouring concrete.

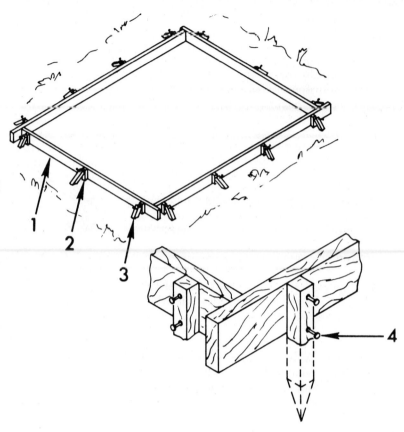

Building Forms

Lengths of lumber [2] used for making forms must be butted tightly together at a stake [3]. This provides a tight, smooth inner surface for the concrete.

Stakes [3] and braces must be driven firmly into the ground. Be sure to drive them straight and true if forms are to be straight up and down.

Cut the stakes [3] and braces just at or below the tops of the forms. Tops of the forms must be unobstructed for operations required in placing and finishing concrete.

Depending on the design requirements of your project, curved forms [4] may be needed. Page 47 describes how to make curved forms.

Expansion joints [1] may be needed where your concrete meets an existing structure. Expansion joints are provided for when you build the forms. Page 48 describes how to build expansion joints.

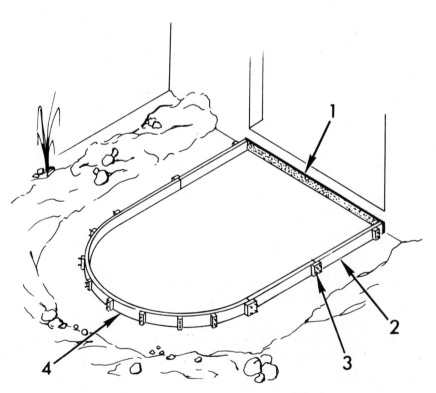

Building Forms

Building forms to allow working with smaller sections [1] of concrete at a time is recommended. The operations required between pouring and finishing concrete must be done rapidly.

When concrete is poured and allowed to sit, surface water will appear. Any operation performed with surface water present can cause chipping and flaking when the concrete hardens. Unless you have plenty of help, you will want to place and finish concrete in smaller sections [1], one at a time.

To allow pouring a continuous length of concrete in small sections, construction joints [3] should be built into the ends of sections which meet. Page 48 describes how to build construction joints.

If your plans call for making control joints [2] in the concrete, now is a good time to mark their positions on your forms. Marks are used as a guide for making control joints during finishing operations.

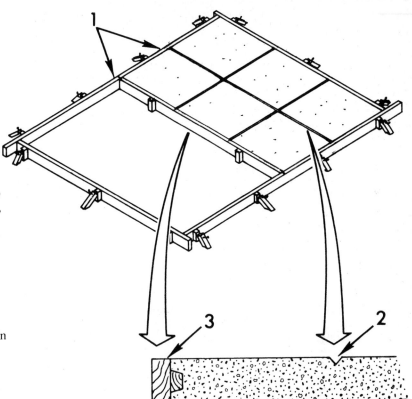

▶ Building Curved Forms

Curved forms [1] can be built with 1-inch thick lumber or 1/4- to 1/2 inch thick plywood, plyscord or hardboard.

Two-inch thick lumber can also be used for building curved forms. However, it requires a series of saw cuts [2] and bending so that the cuts close.

Soaking lumber in water makes bending forms easier. However, soaking plywood is not recommended.

Make the curve as follows:

1. Position the form [4] at its planned slope and curvature, making sure to butt it smoothly and tightly against adjacent forms [3]. Drive stakes [5, 6] on both sides to support the curve.

2. Drive double-headed nails [7] through the outer stakes [6] and into the form [4].

3. Remove the inside stakes [5].

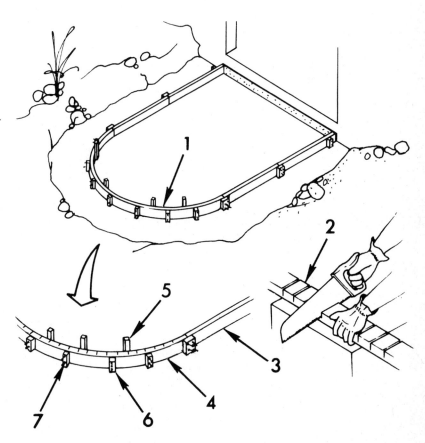

PREPARATION FOR POURED CONCRETE

▶ Making Expansion Joints

When building forms, provisions for expansion joints are made where the new concrete will meet an existing structure. Expansion joints absorb the pressures exerted when the concrete expands.

There are several methods of providing for expansion joints:

- The wood form [1] can remain in place against the structure after concrete is poured.

- A 1/2-inch thick waterproof bituminous filler [2] can be installed in place of the form against the structure. It is generally available in rolls.

- Wedge-shaped lengths of lumber [3] are built and used as the form against the structure. After concrete is allowed to cure, the form is removed. Hot tar is then poured into the space between the concrete and structure.

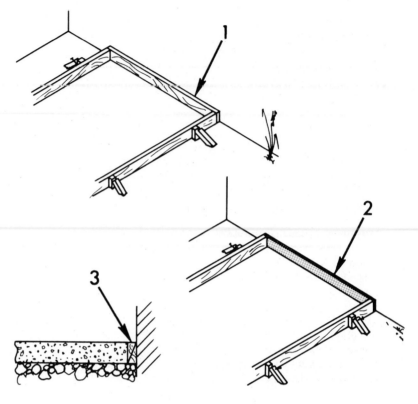

▶ Making Construction Joints

Construction joints [3] are made at end forms to allow placing and finishing small sections of a continuous length of concrete at a time.

The construction joint [3] produces a keyed notch in the previously finished concrete. This provides a firm anchor for new concrete.

Make a construction joint as follows:

1. Taper a board [2] as shown and attach it along the entire length of the end form [1].

2. Place and finish concrete in the first section.

3. Remove the end form [1] when ready to pour the next section.

4. Mix a thin paste of cement and water. Apply the paste over the entire surface of the construction joint [3] immediately before pouring new section. The paste provides a good bond for the new concrete.

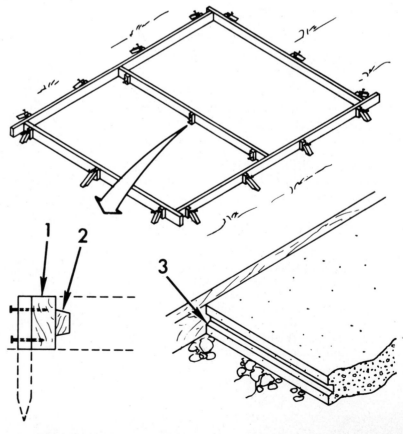

▶ Reinforcing Concrete

Steel reinforcements are used to increase the strength of concrete. Although reinforcement is not always mandatory for your project, it is generally a good idea for most jobs. Building codes should be checked for their reinforcement requirements.

The instructions in this section provide general guidance for using reinforcements. Specific information is described in Design Considerations, Page 32, where applicable.

Reinforcements keep concrete from pulling apart from gravitational or load forces. For most home projects, they are placed midway in the concrete thickness. However, depending on project design, they may be placed wherever the forces are greatest.

Regardless of where reinforcements are positioned within the concrete, at no time should they be within 1 inch of the surface.

Reinforcements should be clean and free of rust for best results. Dirt and rust can prevent a good bond between the reinforcements and the concrete.

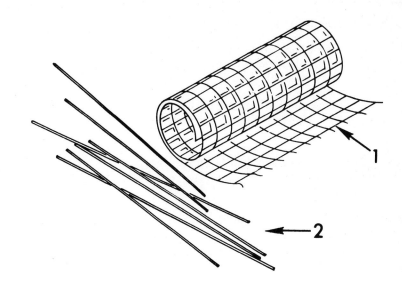

There are two types of reinforcements suitable for most home projects:

Welding wire fabric [1]. See below.

Steel reinforcing rods [2]. Page 50.

▶ Reinforcing with Welding Wire Fabric

Welding wire fabric [1] is used in concrete walls and slabs. It is usually available in rolls of 5- or 6-foot widths.

Position welding wire 2 inches from the forms. Lengths of wire should be overlapped [2] by at least 6 inches and tied together to provide one continuous layer. Use wire cutters for making required cuts.

There are two basic methods for installing welding wire in concrete slabs:

● Place welding wire [1] within forms. Pour concrete as normal. During the spading operation, use a rake to lift wire to desired position in concrete.

● Pour concrete in forms to desired level of welding wire [1]. Place wire over layer of concrete. Place top layer of concrete and finish.

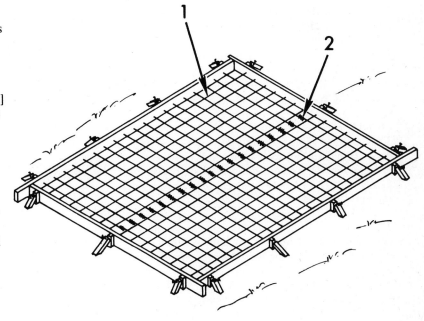

PREPARATION FOR POURED CONCRETE

▶ Reinforcing with Steel Rods

Steel reinforcing rods [1] are used in poured concrete footings and in walls made of all types of masonry materials.

Reinforcing rods are available in lengths up to 20 feet. However, they can be bought precut to your length requirements.

The diameter of the rods range in size. For most home projects, 3/8- or 1/2-inch rods are recommended.

Specific applications of reinforcing rods [1] are described in the appropriate sections of Design Considerations, Page 32. However, you should be familiar with the general guidelines for their use.

When rods [2] are placed horizontally in concrete, they are laid in position as the concrete is poured. Where they are laid end-to-end, rods should overlap by at least 18 inches and be tied together with wire. The overlaps should be staggered.

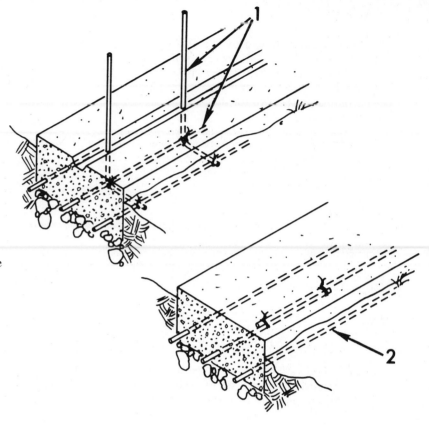

Reinforcing with Steel Rods

Reinforcing rods [6] for walls must be installed when the footing [5] or foundation slab is built. The ends of the rods should be bent 180 or 90 degrees [4] to increase the bond between the rod and the concrete.

The diameter [3] of the 180-degree bend should be six times the diameter of the rod. For the 90-degree bend, extend the rod at least twelve diameters [2]. A length of steel pipe can be used to bend the rods.

The spacing [1] between rods depends on the type of wall being built. For example, when you install rods in a footing for a concrete block wall, space the rods at intervals equal to the core openings of the concrete blocks. Building codes usually describe spacing requirements.

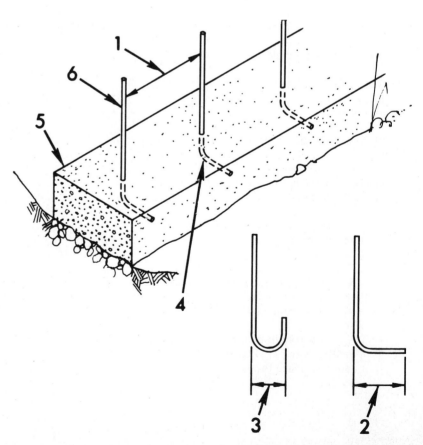

50

Instructions in this section describe how to determine the volume of concrete needed for your project. If you buy concrete materials separately to mix yourself, information on the amount of materials to buy is also provided. Methods of buying concrete are described on Page 52.

Table 1 gives the volume of concrete needed in cubic yards for various concrete thicknesses and surface areas. Table 2 gives the same values in cubic feet. Both tables include a 10% allowance for waste. Use Table 2 if buying concrete materials separately.

If the surface area of your project is not in the tables, add amounts corresponding to several areas. For example, use Table 1 to determine the required number of cubic yards of concrete for a 6-inch thickness with a 270 square foot area:

Partial area of 200 = 4.08 cubic yards
Partial area of 50 = 1.02 cubic yards
Partial area of 20 = .41 cubic yards

Total area of 270 = 5.51 cubic yards

In this case, your project requires 5-1/2 cubic yards of concrete.

If you are buying Portland cement, sand and coarse aggregate separately, go to next section (below) to determine the quantities of each to buy.

Table 1. Cubic Yards of Concrete

Concrete Thickness (inches)	SURFACE AREA (square feet)						
	5	10	20	50	100	200	500
3	.05	.10	.20	.51	1.02	2.04	5.09
4	.07	.14	.27	.68	1.36	2.72	6.79
5	.08	.17	.34	.85	1.70	3.39	8.48
6	.10	.20	.41	1.02	2.04	4.08	10.18

Table 2. Cubic Feet of Concrete

Concrete Thickness (inches)	SURFACE AREA (square feet)						
	5	10	20	50	100	200	500
2	1.4	2.8	5.5	13.8	27.5	55.0	137.5
4	1.8	3.7	7.3	18.3	36.7	73.3	183.3
5	2.3	4.6	9.2	22.9	45.8	91.7	229.2
6	2.8	5.5	11.0	27.5	55.0	110.0	275.0

The amount of concrete materials to buy is affected by the size of coarse aggregate you use. The size of coarse aggregate must be no larger than 1/5 the desired thickness of your concrete.

Table 3 gives the amount of concrete materials needed for 1 cubic foot of concrete.

Multiply the values in Table 3 by the number of cubic feet estimated above.

Table 3. Concrete Materials by Weight in Pounds for 1 Cubic Foot of Concrete

Concrete Materials	Maximum Diameter of Coarse Aggregate	
	3/4 Inches	1-1/2 Inches
Cement	25	40
Sand	44	40
Coarse Aggregate	65	75

There are three basic ways of buying concrete:

- Buy transit-mixed concrete delivered to your home and ready to pour.

- Buy the Portland cement, sand and coarse aggregate separately for you to mix yourself.

- Buy premixed concrete materials in sacks. You have to add only water to mix the concrete.

Transit-mix concrete, sometimes called ready mix, is recommended for larger jobs. It is sold by the cubic yard (27 cubic feet). Normally, your project must require at least 1 cubic yard of concrete before a transit-mix company will deliver.

When you order transit-mix concrete, ask the dealer to deliver the type of concrete you require (air-entraining, early-strength or plain).

When you buy transit-mix concrete, it is essential that you have all the tools, supplies and helpers ready to complete the job at one time. However, transit-mix concrete will save you the work of mixing it, and you will probably get more uniform results.

You will want to buy concrete materials separately or buy premixed concrete in sacks if your job can be done in sections and you want to work at your own pace.

When you buy concrete materials separately, buy Portland cement with early-strength and air-entraining additives already included if your project requires them.

The sand you buy should be washed concrete sand. The coarse aggregate should be washed and graded screened gravel and/or crushed rock. Dirty sand or aggregate results in poor concrete.

━━ **TOOLS AND SUPPLIES** ━━━━━━━━━━━━━━━━━

The following tools and supplies are required for mixing, placing and finishing poured concrete:

- Cement mixer [1]. Mixers can be rented from an equipment rental dealer. Mixing capacities range from 1-1/2 to 3-1/2 cubic feet of concrete.

- Four containers [2], 3 to 5 gallon capacity, for measuring proportions of concrete materials to be mixed.

- Chutes [3] can be built if required for pouring concrete into hard-to-reach areas.

- Contractor-type wheelbarrow [4] with rubber tires. When moving concrete by wheelbarrow, rubber tires help cushion the mix, preventing the coarse aggregate from settling out.

- Square-edged shovels [5] and mortar hoes [6] or garden hoes [7] to spade and push concrete into place.

- Length of unwarped 2 x 4 lumber [8] for striking off concrete to the height of the forms. Length should be at least 18 inches longer than the distance between forms.

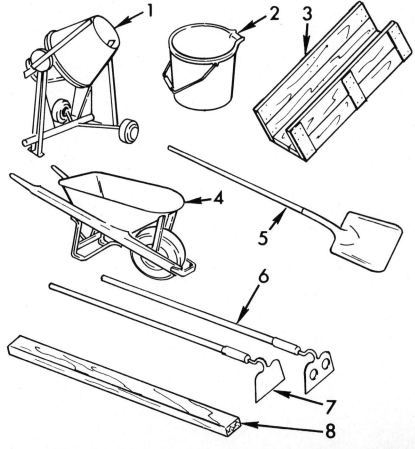

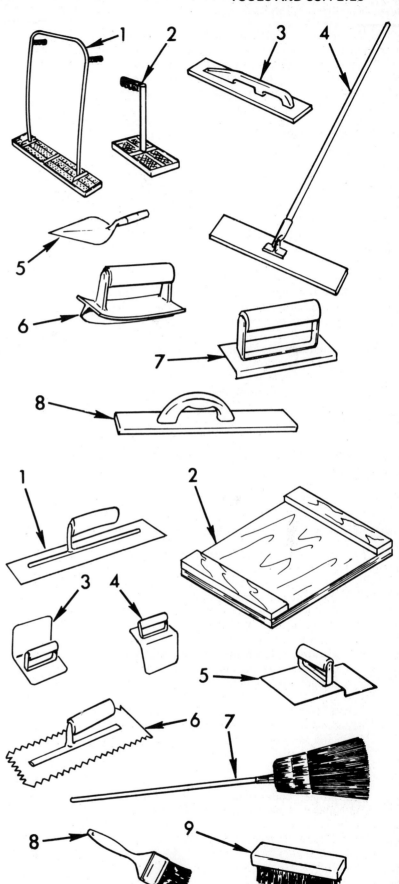

- Concrete tamper [1] to push coarse aggregate below the surface. Hand tamper [2] can be used for smaller jobs.

- Wood or metal darby [3] for initial smoothing of the concrete surface. Wood is recommended for most concrete. Sizes range from 3 to 8 feet.

- Wood or metal bullfloat [4] for reaching areas too large for a darby [3].

- Pointing trowel [5] for cutting wet concrete edges away from forms.

- Groover [6] for making control joints.

- Edger [7] for finishing the edges of the surface.

- Float [8] for compacting the concrete and removing any marks made during previous operations. Floats can be wood or metal. Wood floats are often used as the final finish for a non-slip texture.

- Steel cement trowel [1] for producing a smooth, dense finish.

- Kneeboards [2] can be built for using a float and cement trowel on large surfaces. Kneeboards distribute your weight over the concrete during these operations.

- Special-application tools include inside and outside step tools [3, 4] and driveway approach tool [5]. Step tools can be used only if removing forms immediately after concrete hardens enough to support itself.

- Tools to obtain rough-textured finishes include scratcher trowels [6], brooms [7], soft brushes [8] and wire brushes [9].

- Heavy gloves and rubber boots to protect your skin and clothing from contact with wet concrete.

The period of time between mixing concrete and finishing concrete is relatively short. This section describes the things you must have ready before beginning. Considerations for obtaining the best results with your project are also provided.

Your forms must be built to the requirements of your project. Be sure that provisions for expansion joints and construction joints are made. The location of control joints should be marked on the side forms.

Check that all stakes and braces are below the tops of your forms. Have extra stakes and braces ready in case the forms begin to sag when the concrete is poured.

If you anticipate pouring concrete from a wheelbarrow, build runways up to or over the forms at appropriate locations. Be sure they are strong enough to support the load.

Have any required reinforcements in place or ready for installation.

Decisions on moving and pouring concrete must be made at this time.

If transit-mix concrete is being delivered, consider the following:

- Make provisions for bringing the truck as close to the point of use as possible. The truck chute can be expected to extend up to 17 feet and move in an arc of nearly 180 degrees.

- Have chutes built and ready if the truck chute will be unable to reach your project.

- A loaded truck can damage curbs, driveways, sidewalks and your lawn. Use blocks at the curbs and heavy planks to distribute the weight of the truck.

- Consider the height and width of the truck to prevent damage to trees and structures around your home.

If you mix the concrete yourself, consider the following:

- Mix the concrete as close as possible to the point of use. For concrete materials purchased separately, have the sand, coarse aggregate and cement delivered near where they will be mixed.

- For jobs which can be completed in small sections, a wheelbarrow-type mixer can be used. You can mix concrete and move it to the point of use in the mixer.

Have all the required tools and supplies ready. Be sure all materials for making your desired finish and for curing concrete are handy and ready for use.

Have plenty of help. Depending on the size of your project, you may be able to complete the job yourself. However, at least one helper is recommended for any large project.

Before you begin placing concrete, prepare the base and forms as follows:

- Dampen the base to prevent water from being drawn out of the concrete mix. Do not allow puddles to form.

- Dampen the forms to seal any cracks in the lumber. A light coat of motor oil can be applied to make removing the forms easier. Do not apply oil if pouring colored concrete.

If pouring concrete against finished concrete which contains a construction joint, mix a thin paste of Portland cement and water. Apply the paste to the joint to obtain a bond for the new concrete. Be sure concrete is placed before the paste dries.

Instructions in this section describe how to mix the correct proportions of separate concrete materials. Using a cement mixer and mixing by hand are included.

Be sure to read Things to Have Ready Before Mixing and Placing Concrete, Page 54, before continuing. The time from mixing concrete to that of completing the finishing operations is relatively small. You will want to have all the tools, materials and helpers ready to complete the work.

If using premixed concrete in sacks, the information on mixing in this section is generally applicable. However, follow manufacturer's instructions for specific directions.

There are two methods of obtaining the correct proportions of Portland cement, sand, coarse aggregate and water:

- Measuring proportions by weight. See section below.

- Measuring proportions by volume. Page 56.

The preferred method is measuring by weight. It ensures more uniform results and allows for accurate adjustment of proportions. Proportions may need adjusting depending on the moisture content of the sand and the consistency of your first concrete mixture.

▶ **Measuring Proportions by Weight**

The table lists the correct proportions of concrete materials by weight for 1 cubic foot of concrete. Values depend on the maximum size of your coarse aggregate.

The moisture content of your sand must be checked before measuring. Page 56.

Each material is weighed in separate 3 to 5 gallon containers. A bathroom scale can be used for weighing. Be sure to make adjustments for the weight of each container.

After weighing each material, mix your first batch of concrete. Mixing instructions begin on Page 57. Make any required adjustments to your mixture after checking its consistency.

When the correct proportions are obtained, mark the level of material in each container for future batches. You will not need to use the scale again.

Concrete Materials by Weight in Pounds for 1 Cubic Foot of Concrete

Concrete Materials	Maximum Diameter of Coarse Aggregate	
	3/4 Inches	1-1/2 Inches
Cement	25	23
Wet Sand	44	40
Coarse Aggregate	65	75
Water	10	9

▶ Measuring Proportions by Volume

The table lists the correct proportions of concrete materials by volume. Values depend on the maximum size of your coarse aggregate.

The moisture content of your sand must be checked before measuring. See section below.

Measure out the listed amount of materials, using the same size container for each. For example, if using 3/4-inch aggregate, measure 1 pail of cement, 2-1/4 pails of sand, 2-1/2 pails of aggregate and 1/2 pail of water.

Mix your first batch of concrete. Mixing instructions begin on Page 57. Make any required adjustments to your mixture after checking its consistency.

When the correct proportions are obtained, note the amounts used of each material to aid in mixing future batches.

Concrete Materials by Proportions of Volume

Concrete Materials	Maximum Diameter of Coarse Aggregate	
	3/4 Inches	1-1/2 Inches
Cement	1	1
Wet Sand	2-1/4	2-1/4
Coarse Aggregate	2-1/2	3
Water	1/2	1/2

▶ Checking the Moisture Content of Sand

The proportions listed in the tables on Pages 55 and above for mixing concrete are based on the use of wet sand. The moisture content of your sand must be checked to see if adjustments must be made to the amount of water added.

Accurate adjustments can only be made if you are measuring materials by weight. If measuring by volume, accurate adjustments are hard to make. Try to use wet sand when measuring by volume.

Check your sand as follows:

Squeeze some sand in your hand and try to form it into a ball.

- Wet sand forms a ball, leaving no water on your hand. No adjustment is required.

- Damp sand will not form a ball. It crumbles very easily. Increase the amount of water in your mixture by 1 pound for each cubic foot of concrete.

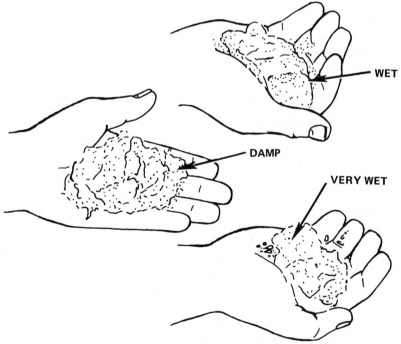

WET

DAMP

VERY WET

- Very wet sand forms a ball, but leaves free water on your hand. Decrease the amount of water by 1 pound for each cubic feet of concrete.

Checking the Moisture Content of Sand

After determining the correct proportions of concrete materials to use, you are ready to mix concrete.

WARNING

Wet concrete is caustic to your skin. Be sure to wear protective gloves and boots when working with concrete mix.

Wet concrete must not be allowed to dry out before you begin placing it. If it becomes too hard to work with, you may try remixing it.

CAUTION

If remixing a batch of concrete, do not add water. Water will destroy the correct proportions of materials. The batch must be scrapped.

There are two methods of mixing concrete:

Mixing with a Cement Mixer. See next section.

Mixing by Hand. See next section.

▶ Mixing with a Cement Mixer

The first batch of concrete mixed is partially used to check its consistency. Any required adjustment to the proportion of your materials can then be made.

When you have the desired proportions with an acceptable consistency, record the proportions for future batches.

1. With the cement mixer stopped, add one-half the water and all of the coarse aggregate. Start the mixer.

2. Add all of the sand, all of the cement, and the remaining water. Allow the mixer to operate for a minimum of 3 minutes. The mixture must have a uniform color throughout.

If mixing the first batch of concrete, go to Page 58 to check its consistency. Otherwise, you are ready to place and finish the concrete. Instructions for placing concrete begin on Page 58.

▶ Mixing by Hand

Mixing by hand can be done in a mortar box, a wheelbarrow or on a flat concrete surface.

The first batch of concrete mixed is partially used to check its consistency. Any required adjustment to the proportion of your materials can then be made.

When you have the desired proportions with an acceptable consistency, record the proportions for future batches.

1. Mix all of the cement and sand together until it has a uniform color.

2. Thoroughly mix all of the coarse aggregate into the cement-sand mixture.

3. Make a bowl-shaped cavity [1] in the center of the mixture.

After the water is poured into the cavity [1], concrete is mixed by pulling the materials up from the outer edges into the water. Use a hoe for mixing.

Work carefully. Do not allow the water to run out of the cavity [1]. Mix materials until all of

the coarse aggregate is covered completely with the cement-sand-water mixture.

4. Add all of the water into the cavity [1]. Mix the concrete.

If mixing the first batch of concrete, go to Page 58 to check its consistency. Otherwise, you are ready to place and finish the concrete. Instructions for placing concrete begin on Page 58 (bottom).

MEASURING AND MIXING CONCRETE

► **Checking the Consistency of Concrete Mix**

Check the consistency of the first batch of concrete mix as follows. Be sure to note any adjustment to the proportion of concrete materials for future batches.

1. Check that the mixture slides off your trowel or shovel. It must not flow off.

2. Drop a shovelful of mixture on the ground. Check that it tends to stay together. It must not spread out flat.

If the mixture fails Steps 1 or 2, add equal proportions of sand and coarse aggregate.

3. Make a mound of concrete. Lightly drag a trowel across the surface of the mound.

4. Check that the mixture becomes smooth on the surface with no spaces between the coarse aggregate.

If the mixture fails Step 4, add 1 part water to 2 parts cement.

PLACING CONCRETE ═══════════════════════

After the concrete is mixed, or the transit-mix concrete arrives, you are ready to place it in the forms and finish it. Placing concrete consists of the following operations:

    Pouring and distributing
    Spading
    Striking off and tamping
    Bullfloating and darbying

When darbying is complete, you must wait a period of time before beginning the finishing operations. The waiting period varies depending on weather conditions and how fast the concrete hardens. Finishing operations begin on Page 61.

CAUTION

Be sure to perform the pouring through darbying operations before water from the mix collects on the surface. Any operation with surface water present can cause the concrete to chip and flake when it hardens.

► **Pouring and Distributing**

Concrete is poured and distributed uniformly in depth within the forms. Pour concrete toward concrete already in place.

When pouring concrete, do not drop it any farther than necessary. You do not want to jar the mixture.

Use a hoe or square-ended shovel to distribute the poured concrete. Push the concrete into position. Do not pull it toward you.

When the concrete is uniformly distributed and noticeably above the height of the forms, go immediately to the spading operation. Page 59.

▶ **Spading**

Spading ensures that the concrete is uniformly compacted within the forms.

Use a shovel [1] to dig into the mixture, especially at corners, around the form edges and around any reinforcements. Be sure all spaces within the forms are filled.

When spading is completed, go immediately to the striking off and tamping operations.

▶ **Striking Off and Tamping**

Striking off and tamping are done to level the mixture at the height of the forms and to push the coarse aggregate slightly below the surface.

These operations can be done in either order, although striking off first is recommended.

Use a length of 2 x 4 lumber [2] longer than the distance between the forms to strike off the concrete. The 2 x 4 is moved rapidly side to side in a sawing motion, using the tops of the forms as a leveling guide.

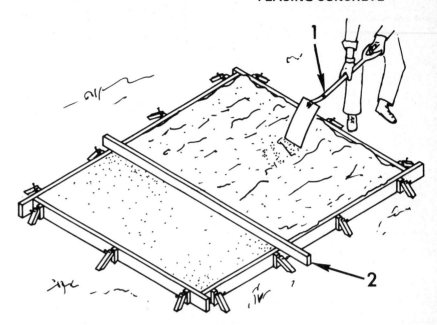

Slightly tilt the 2 x 4 to form a cutting edge. Push excess concrete ahead of it, filling low spots along the surface. Repeat the process in the opposite direction until concrete is level with the tops of the forms over the entire surface.

Striking Off and Tamping

After you strike off the concrete, it should be immediately tamped. Use a tamper [1] to push the concrete into any remaining spaces and to push the coarse aggregate slightly below the surface.

CAUTION

Do not overdo the tamping operation. If coarse aggregate is pushed too far below the surface, excess sand-cement mixture will be on top. This may result in unequal shrinkage during curing and cause cracking.

Work the tamper in rows along the surface. Walk backwards away from the tamping operation.

When tamping is completed, go immediately to the bullfloating and darbying operation. Page 60.

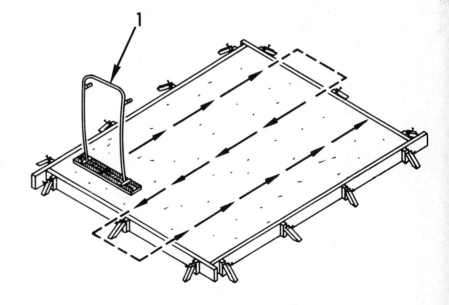

PLACING CONCRETE

▶ Bullfloating and Darbying

Bullfloating and darbying achieve final leveling and smoothing of the surface. Use a bullfloat [1] for areas which cannot be reached with your darby [2].

CAUTION

Do not overwork the concrete. Excessive bullfloating and darbying produce a less durable surface.

A darby [2] is used by holding it flat against the surface while moving it side to side in a sawing motion. The blade of the darby levels the surface by cutting off bumps and filling low spots.

After leveling, slightly raise the leading edge of the darby to smooth the concrete.

A bullfloat [1] obtains the same results as a darby. Push the bullfloat away from you with its leading edge slightly raised. Pull it back with the blade flat to cut off bumps and fill any low spots.

If low spots remain in the surface, shovel in additional concrete mix and reuse your darby or bullfloat.

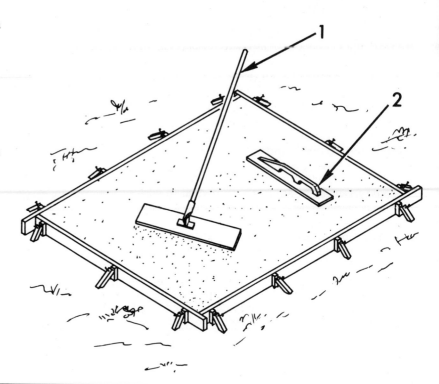

Bullfloating and Darbying

Immediately after darbying, use a pointed trowel [1] to cut the concrete away from the forms [2]. Depth of the cut should be about 1 inch.

When the surface is level and completely smooth, and the concrete is cut away from the forms, you must wait for a period of time before beginning your desired finishing operations.

On hot, dry days, there may be no waiting period. On cool, humid days, the waiting period may be as long as several hours.

Wait until all surface water disappears. If using air-entraining concrete, no surface water may be visible.

Regardless of the type of concrete, wait until the surface hardens enough to where foot pressure leaves a maximum indention of 1/4-inch in the concrete.

Go to Page 61 for finishing your concrete.

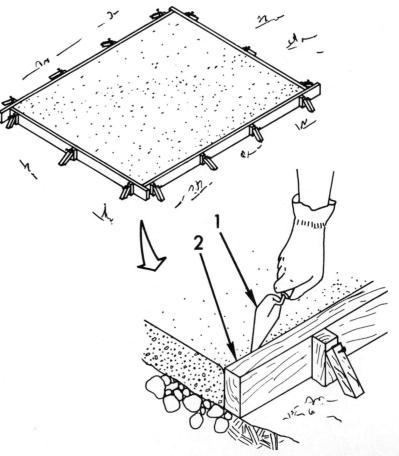

The techniques for finishing concrete vary with the type of finish you plan. Nearly all types are modifications of producing a plain concrete finish.

The instructions in this section describe how to make a plain concrete finish. The different types of finishes and the methods of obtaining them are described beginning on Page 28.

Finishing concrete consists of the following operations:

> Edging
> Making control joints
> Floating
> Troweling

When final troweling is complete, you must cure the concrete. Instructions for curing concrete are given on Page 64.

Go to next section (below) to begin edging.

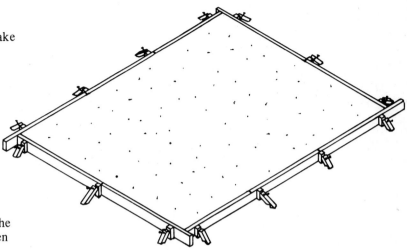

▶ Edging

Concrete surfaces require edging so that when forms are removed, the edges of the surface have a smooth, finished appearance. An edger [1] is used for this operation.

Edging and making control joints can be done at the same time. Making control joints is discussed on Page 62.

CAUTION

Do not apply too much pressure when using an edger [1]. Indentions left by the tool may be hard to remove during the floating and troweling operations.

Edges are finished by placing the edger [1] flat against the surface using the forms [2] as a guide.

Move the edger [1] first in one direction, then the other. Slightly tilt the leading edge to prevent gouging the concrete.

When all edges have been finished, go to Page 62 (top) to make control joints. If no control joints are required, go to Page 62 (bottom) for floating.

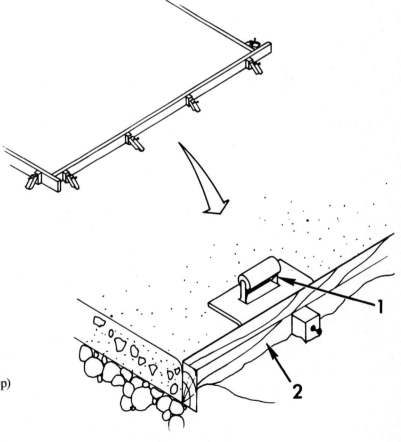

FINISHING CONCRETE

▶ Making Control Joints

Control joints [1] are required to prevent cracks in the surface after it cures. A groover [4] is used to make control joints. Joints extend from 1/5 to 1/4 the thickness of the concrete.

A straight 1-inch board [2] longer than the distance between the forms can be used as a guide for using the groover. Use the marks [3] made on your forms when they were built as a guide for the position of control joints.

CAUTION

Do not apply too much pressure when using a groover [4]. Indentions left by the tool may be hard to remove during the floating and troweling operations.

To start the joint, push the groover [4] into the concrete. While applying pressure to the back of the tool move it forward along the line of the joint. Make the joint first in one direction, then the other.

When all control joints [1] are made, go to next section (below) to begin the floating operation.

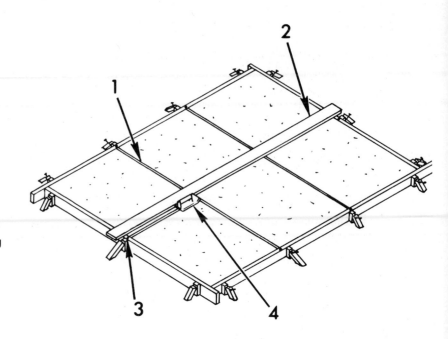

▶ Floating

After edging and making any required control joints [1], you are ready to float the surface.

Floating compacts the surface and brings the sand-cement mix [2] to the top in preparation for any desired finishes. Floating is often done as a final finish. It also removes any imperfections left in the surface by previous operations.

CAUTION

Do not begin floating if it brings water to the surface. This can cause the concrete to flake after it hardens. However, do not wait until the surface is too hard to finish.

When making a plain finish, floating and troweling are usually done on one area at a time. Page 63 describes troweling. Do not trowel an area, however, without first using a float.

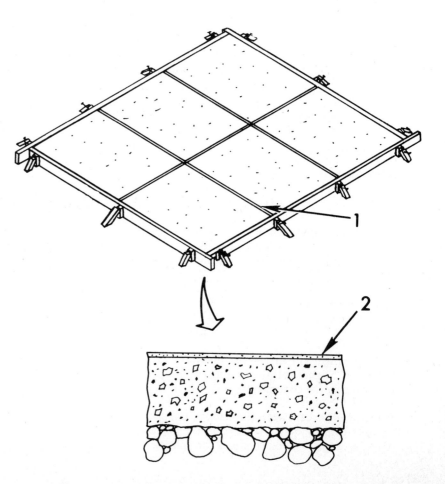

Floating

For large areas of concrete, use kneeboards [1] to support yourself. If floating and troweling at the same time, complete these operations before moving to another area.

Wood or metal floats [2] may be used. Wood floats produce a rougher finish and should be used if floating is the final operation.

Metal floats [2] provide a smoother finish. Use a metal float if further finishing operations are required.

A float [2] is used by holding the blade flat against the surface. Move it in wide, sweeping arcs with a slight sawing motion.

After completing the floating operation, go to next section (below) if troweling is required. If floating is the last operation, go to Page 64 for instructions on curing concrete.

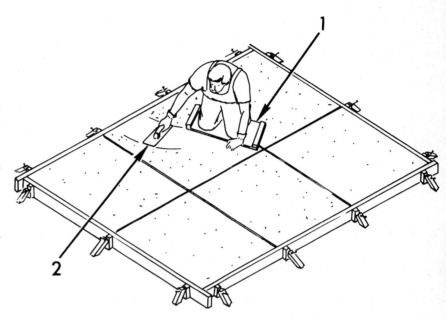

▶ Troweling

Troweling is the last operation of producing a plain concrete finish. It achieves a dense, hard, smooth surface.

CAUTION

Do not trowel the surface without first performing the floating operation. Page 62.

You can trowel the surface as many as two or three times. Each time a smoother, harder surface is provided.

For the first troweling, hold the trowel [1] flat against the surface. Move it in wide sweeping arcs with each arc overlapping the previous one by about one-half.

Wait a few minutes before troweling a second time to allow the surface to harden. Begin the second troweling when hand pressure produces only a slight indention.

Hold the trowel [1] with its leading edge slightly tilted. Use the same motion as for the first troweling until the desired finish is produced.

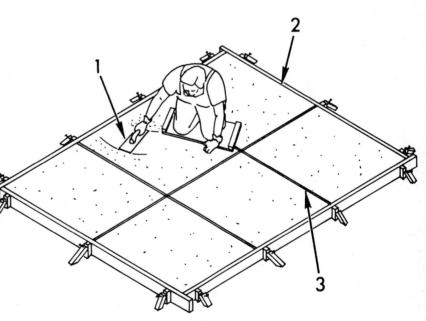

When troweling is complete, inspect the edges [2] of the surface and any control joints [3]. Perform the edging and making control joint operations again if required.

Go to Page 64 for instructions on curing concrete.

Concrete must be kept warm and moist after it is finished to allow for maximum hardness and durability. The strength of concrete increases under these conditions as it hardens.

For best results, concrete should be allowed to cure for a full week. Do not let concrete dry out during this time.

There are several good methods for curing concrete:

- Ponding. Build a dam around the edges of your concrete. Keep a layer of water over the entire surface.

- Spread burlap over the concrete. Keep the burlap wet.

- Commercial curing agents are available. Follow manufacturer's instructions for the method of application.

- Cover the entire surface with plastic sheeting. Keep the plastic absolutely air tight and smooth. If the sheeting does not lay smooth, the concrete may discolor.

- Use a hose to spray the concrete with water several times a day.

After the curing period, your forms can be removed.

<u>CAUTION</u>

Be sure to remove forms very carefully. Avoid chipping the concrete and damaging the form lumber.

If the concrete is damaged or cracked when the forms are removed, it should be repaired as soon as possible.

■■ CONCRETE REPAIRS ■■■

▶ **Repairing Hairline Cracks**

The following tools and supplies are required:

Putty knife [1]
Portland cement. Because patches generally appear darker than the repaired surface, use a proportion of white Portland cement if coloring is important.
Water

1. Moisten crack [2] thoroughly. Keeping crack moist overnight is best.

2. Mix a thick paste of Portland cement and water.

Crack [2] must contain no standing water when repair is made. All loose particles must be removed.

3. Using putty knife, fill crack [2] completely with paste. Smooth surface as required.

4. Allow repaired area to cure for several days by keeping it moist.

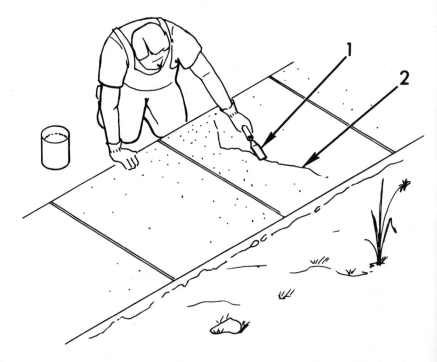

▶ Repairing Large Cracks

The following tools and supplies are required:

Cold chisel [1]	Putty knife [4]
Hammer [2]	Paint brush [5]
Wire brush [3]	

Portland cement
Epoxy or latex cement patching compound
Water

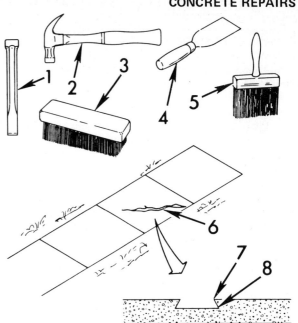

1. Using hammer and chisel, remove concrete beneath edges [7] of crack until crack is wider at bottom [8] than at edges.

2. Moisten crack [6] thoroughly. Keeping crack moist overnight is best.

Crack [6] must contain no standing water when repair is made. All loose particles must be removed.

3. Mix a paste of Portland cement and water. Apply paste to crack [6].

Cement paste must not be allowed to dry before applying patching compound.

4. Using putty knife, fill crack [6] completely with patching compound. Smooth surface as required.

5. Allow repaired area to cure according to instructions of the patching compound.

▶ Repairing Broken Edges and Corners

The method of repairing broken edges and corners [4] depends on whether or not you have the old piece of concrete [5]. If the old piece is available, cement it into place using epoxy or latex cement.

If the old piece of concrete [5] is not available, you must patch the break with patching compound or with a thick mix of Portland cement, sand and water.

The following tools and supplies are also required:

Wire brush [1]
Putty knife [2] or trowel [3]

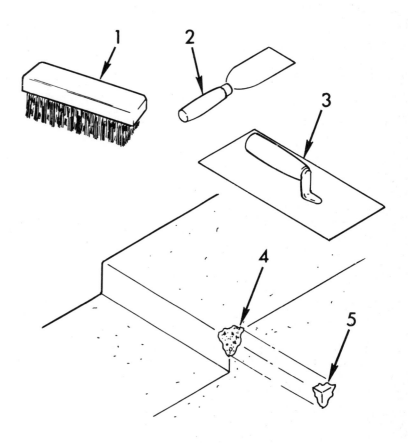

1. Moisten break [4] thoroughly. Keeping break moist overnight is best.

Break [4] must contain no standing water when repair is made. All loose particles must be removed.

2. Mix a thin paste of Portland cement and water. Apply paste to break [4].

Cement paste must not be allowed to dry before repairing break [4].

CONCRETE REPAIRS

Repairing Broken Edges and Corners

If replacing old piece of concrete [2], go to
Step 5. If patching break [1] with new concrete,
continue.

3. Mix a thick paste of Portland cement, sand
   and water. Using the paste, build up
   broken edge or corner [3].

4. Allow paste to harden slightly. Smooth
   and finish edge or corner [3] to match
   surrounding area. Go to Step 7.

5. Apply epoxy or latex cement to old piece
   of concrete [2]. Place and hold piece at
   installed position.

Concrete [2] may need support to hold it in
place until cement hardens. A board can be
used for this purpose.

6. When concrete [2] can support itself, wipe
   off excess cement.

7. Allow repaired area to cure for several days
   by keeping it moist.

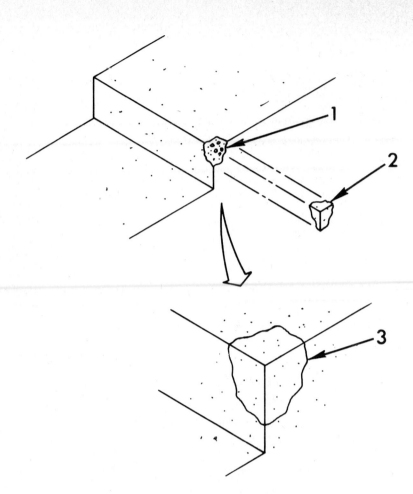

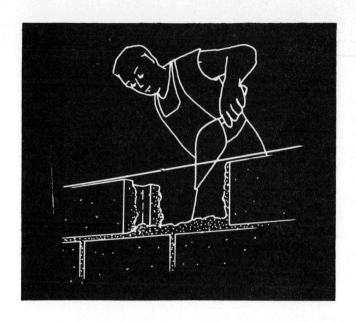

CONCRETE BLOCK WORK

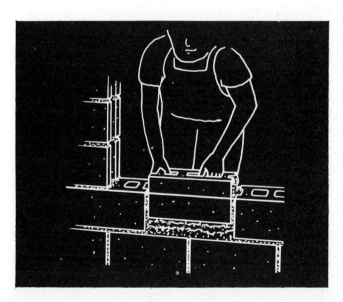

Before you begin any concrete block construction, it is a good idea to first become familiar with:

- Local building codes and construction ordinances.

- The minimum property or construction standards of the Federal Housing Authority (FHA) and the Veterans Administration (VA).

Your new construction must comply with the FHA and VA requirements if these agencies are to guarantee future loans on the property.

For many kinds of construction, building permits are required. Building specifications and regulations are available from the agencies that issue these permits. In many cases, inspections may be required during the course of the job to ensure that the construction conforms to regulations.

Building and construction regulations are likely to provide specific directions on such items as:

- Retaining Walls
  Footing Depth
  Concrete Mix for Footings
  Grout Mix
  Mortar Mix
  Concrete Block Units
  Reinforcements
  Type of Design Requiring Permits
  Type of Design Not Requiring Permits

- Home Masonry Walls
  Footings
  Exterior Walls
  Interior Load Bearing Walls
  Interior Non-Load-Bearing Walls
  Solid Masonry
  Hollow Unit Masonry

The instructions in this section represent good construction practice. However, they do not necessarily conform to all the codes and regulations required for your particular area. Therefore, be sure to check with your local agencies before beginning a construction job.

▶ **Types and Uses of Concrete Blocks**

Concrete blocks can be produced from a variety of materials. In addition to the most commonly used aggregate materials such as sand and gravel or sand and crushed rock, materials such as cinders, pumice and shale are used.

Although sizes and shapes of concrete blocks vary, the most common construction block used is the 8 in. x 8 in. x 16 in. block with hollow cores [1]. Actual dimensions of this block are 7-5/8 in. x 7-5/8 in. x 15-5/8 in. The 3/8-inch difference allows for the mortar joint. Therefore, when planning and estimating materials for a job, remember that the block dimensions already make allowance for the thickness of mortar joints.

Other blocks [2] used for making walls may be:

- 4 or 8 inches wide
- 4, 8 or 12 inches high
- 8 or 16 inches long

Blocks that are 12 inches high are used for making foundation walls.

Blocks that are 4 inches wide are used for making partitions and cavity walls.

Blocks that are 8 inches wide can be used for almost any kind of wall.

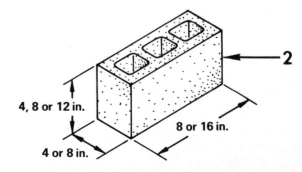

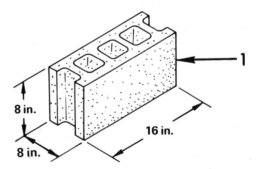

Types and Uses of Concrete Blocks

Corner blocks [1] are smooth on one end and are used at the ends of walls.

Stretcher blocks [2] are the blocks used between the corners of walls. They are smooth on one face.

Solid blocks [3], or capping blocks, may be partially cored. They are used to cover the interior cavities of the wall.

Header blocks [4] have a cutout on one side. They are used across the top and bottom of openings such as doors or windows.

Lintels [5] are concrete beams that span the top of openings such as doors and windows.

For modular doors and windows, pre-cast lintels are available made of reinforced concrete.

Lintels may be also constructed on the job from header blocks.

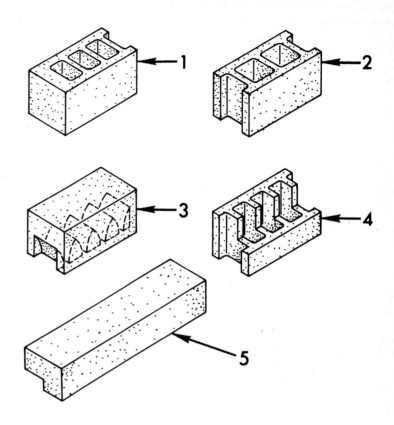

Types and Uses of Concrete Blocks

Jamb blocks [1] have a cutout on one side. They are used at the sides of openings in walls where additional construction is required.

Control joint blocks [2] provide support to wall sections on each side of a control joint. They have tongue-and-groove shaped ends.

▶ Estimating Blocks

The area of one concrete block (8 in. x 16 in.) is 0.878 square feet.

The number of blocks required to make your wall will be the area of your wall (length x height) divided by 0.878.

There are approximately 113 blocks per 100 square feet.

A corner block will be required for each end of each course. Divide the height of the wall (in inches) by 4 (if 8-inch high blocks are to be used). This is the number of corner blocks which are needed.

Estimate special-purpose blocks from job plans.

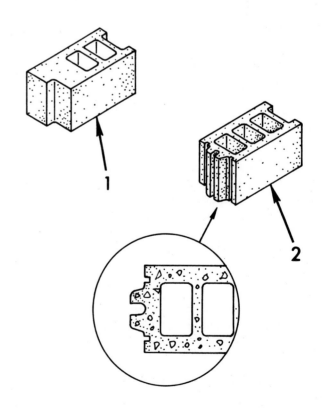

CONCRETE BLOCK WORK

▶ Tools

The following tools are required to construct a block wall:

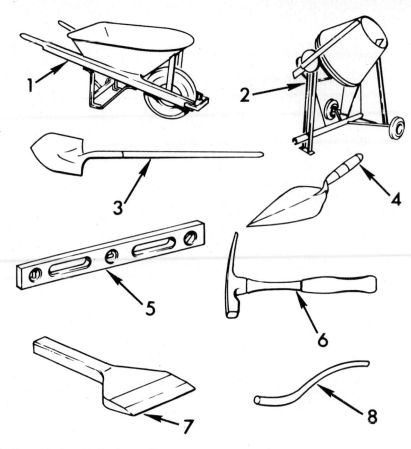

- A wheelbarrow [1].

- A cement mixer [2] if required, to mix mortar in large quantities.

- A shovel [3].

- A trowel [4]. Select a trowel with a durable handle so that it can be used to tap the blocks into final position in the mortar.

- A mason's level [5] to check that the blocks and the wall are level and plumb. A mason's level should be long enough to span more than one block.

- A bricklayer's hammer [6] and chisel [7] to cut and trim blocks.

- A mason's jointer [8] to compact the mortar into the mortar joint. Compacting is required to make mortar joints weather-proof and uniform in appearance.

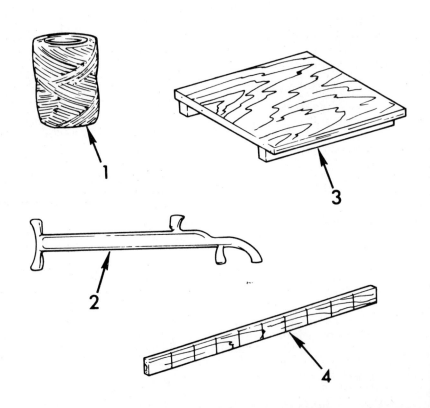

Tools

- A course string [1] to mark the top level position of the row of blocks. A row of blocks is called a course. Any stout cord can serve as a course string.

- Mason's line stretchers [2] to hold the course string in position. Two stretchers are required for each wall.

- Mortar board [3] to transport small quantities of mortar from the mortar mixing site to the place where the mortar is to be applied to the blocks or wall.

- A story board [4]. This is a straight board marked at 4- or 8-inch intervals. It is used to check the height of each row above the foundation.

Your construction job will have some or all of the features shown in the illustration. Procedures for doing each operation are given in the following pages. Some features, such as control joints [1], may be required by local regulation. The details of constructing a control joint may also be specified in the regulation.

Procedures are given for:

- Constructing Control Joints [1], Page 83.

- Laying Lintels [2], Page 82.

- Grouting [3], Page 85.

- Building Up Corners and Ends [4], Page 78.

- Intersecting Walls [5], Page 78.

- Parging [6], Page 87.

- Finishing Mortar Joints [7], Page 86.

- Laying the First Course [8], Page 75.

- Building Up Walls [9], Page 80.

- Laying the Top Course [10], Page 82.

These are different construction details for different types of walls. They are discussed as follows:

Making Structural Walls, next section.

Making Facing Walls, Page 73.

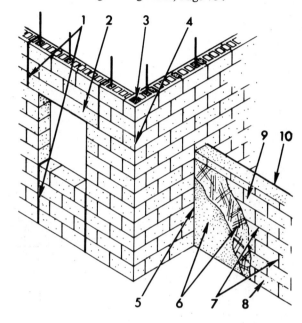

► Making Structural Walls

Concrete blocks may be used to make structural walls. Structural walls are:

- Retaining Walls — where one face of the wall is below ground level, the other face is above ground level.

- Load-Bearing Walls — where the wall is used to support additional construction.

- Basement Walls — Foundation walls that combine the requirements of retaining walls and load bearing walls.

Structural walls can be expected to be subject to building codes, ordinances and minimum standards of FHA and VA requirements.

Some structural walls will require reinforcement rods [1] embedded in the footing and continuing up the wall.

The spacing of the reinforcement rods will be specified in your building permit. Be sure that the spacing is also compatible with the core spacing of the blocks.

Cores that require reinforcements usually must be grouted [3]. Cores may be grouted after several courses are made, or after the wall is completed. Check your building permit. Inspection may be required after wall is completed but before the cores are grouted.

Height control [2] is very important when making load-bearing walls. The top surface of the wall must be horizontally level so that the structure above it is also level and the load on the wall is evenly distributed.

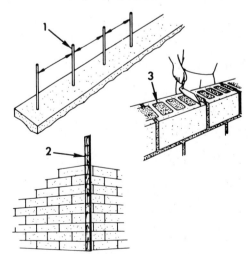

LAYING CONCRETE BLOCKS

Making Structural Walls

Structural walls built below ground level must have adequate drainage provisions.

Drainage provisions for walls may be provided by placing drain tiles to permit removal of seepage.

Sometimes when making retaining walls, for example, proper drainage may be provided by leaving all or part of the vertical mortar joint [1] of the first course open so that moisture may seep through the wall.

Holes in the mortar bed of the first course [2] can accomplish the same thing.

These kinds of drainage provisions are called "weep holes".

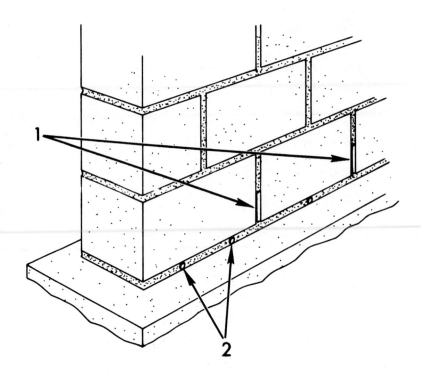

Making Structural Walls

Structural walls where sound or temperature insulation is important may be made using cavity walls.

Cavity walls are parallel walls, usually made of partition blocks, with an air space about 2 inches wide separating the walls.

Cavity walls may be bonded together using corrosion-resistant metal tie bars [1] embedded in the mortar bed at horizontal intervals not more than 32 inches, and vertical intervals not more than 16 inches.

If moisture penetration is possible, drain provisions must be made to remove the moisture from the cavity.

Flashing [2] may be used to direct the moisture away from the inner wall.

Weep holes [3] may be used in the outer wall to permit moisture to escape.

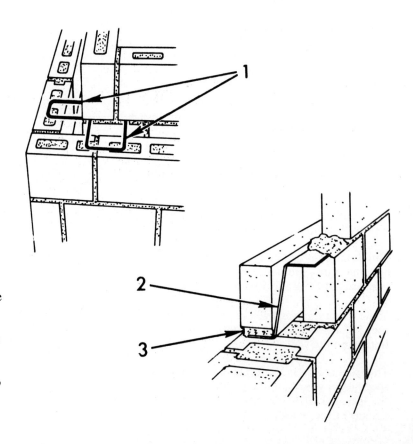

▶ **Making Facing Walls**

Concrete blocks are often used as backup walls to be faced with brick or stone.

When making brick facing on a block backup wall, it is recommended that the walls be built up concurrently.

The block wall may be built up one or two courses ahead of the bricks, or the brick side may be built up a few courses ahead of the blocks.

If the brick side is built up ahead of the block side, the block side of the bricks must be parged [1] before building up the block side.

If the block side is built up ahead of the brick side, the brick side of the blocks must be parged [2] before building up the brick side.

A discussion of parging is found in Mixing Mortar, Page 74, and in Parging, Page 87.

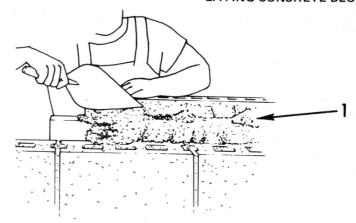

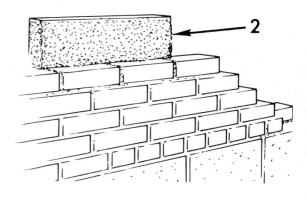

Making Facing Walls

All of the bricks must be laid in a full mortar bed [2].

The brick facing is bonded to the backup wall by a header row of bricks.

A header row [1] of bricks is a row of bricks laid so that the ends are flush with the face of the wall.

Backup walls using blocks 8 inches thick will result in a wall 12 inches thick.

When making a backup wall using 8-inch thick blocks, header row may be either the 6th or 7th row.

If the header row is every 7th row, cement bricks may be used to adjust the spacing of the blocks.

If the header row is every 6th row, header blocks are used in alternate courses.

Backup walls using blocks 4 inches thick will result in a wall 8 inches thick.

When making a backup wall using 4-inch thick blocks, every 6th row of bricks must be a header row.

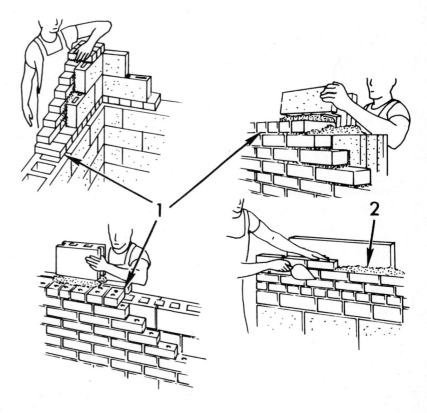

LAYING CONCRETE BLOCKS

▶ Mixing Mortar

Mortar is a mixture of Portland cement, masonry cement, hydrated lime, sand and water. It is used as a bonding agent for masonry materials.

Mortar may also be used to weatherproof a wall by making a plaster-like coating on one face of the wall. This plaster-like coating is called parging.

A thin mortar mix used to fill the hollow cavities of concrete block walls is called grout. Grout may be made using the regular mortar mix, or using a pea gravel/sand mixture as an aggregate.

Mortar is made by mixing ingredients by volume. The proportions vary for different kinds of masonry. Approximately 8.5 cubic feet of mortar will be required for 100 square feet of wall (about 113 blocks).

Some accepted mixing proportions are listed in Table 2. To use the table:

- Select the type of mortar required for your application from Table 1.

- Table 2 prescribes the proper proportions of various ingredients for the different types of mortar.

Table 1. Mortar Uses

Mortar Uses Masonry Item	Type of Mortar Required
Foundations, Walls or Piers	M or S
Exterior Walls Above Grade	S or N
Retaining Walls	M
Partitions	S or N
Reinforced Masonry	M or S
Exterior Cavity Walls	M or S

Table 2. Mortar Mix By Volume

Mortar Mix by Volume Mortar Type	Portland Cement	Hydrated Lime or Lime Putty	Type II Masonry Cement	Damp Loose Mortar Sand
M High Strength	1	1/4	–	3 to 3¼
	1	–	1	4½ to 6
N Medium Strength	1	½ to 1	–	4 to 6
	–	–	1	2¼ to 3
S High Bond	1	¼ to ½	–	3¼ to 4½
	½	–	1	3¼ to 4½
Parging	1	–	–	2½
Grout	1	–	–	2 Sand + 3 Pea Gravel

Mixing Mortar

Mortar must be used within 2-1/2 hours (3-1/2 hours in cool weather) after initial mixing.

Mortar not used within these time limits should be discarded and a fresh batch of mortar mixed.

Only mix the amount of mortar that you will be able to use within the time limits.

If mortar becomes too dry, due to evaporation, within the time limits, additional water may be mixed with the mortar to bring it back to the proper consistency.

Do not remix mortar more than 2-1/2 hours after it was originally mixed.

Mix additional amounts of mortar only when you are ready to use it.

Mix or stir the mortar on your mortar board frequently. This keeps the mixture consistent throughout.

Mortar may be mixed by hand or using a cement mixer.

1. Combine dry ingredients determined from tables above in a mixing container. Thoroughly mix dry ingredients together until they are mixed evenly.

The proper water content is important to permit mortar to be easily applied to the blocks and have good bonding strength. If mortar is too wet, it will not retain shape or position on tools or blocks. If mortar is too dry, it will not adhere properly to the blocks.

The exact amount of water to use cannot be specified. As a rule, add as much water as you can without impairing workability of the mortar. Mortar has good workability if it:

- Spreads easily.

- Clings to vertical surfaces and the underside of horizontal surfaces.

- Does not flow out of the joints between the blocks.

Use drinking-quality water only for best results.

2. Add water and continue mixing ingredients together until mortar has good workability. Mix ingredients for at least 3 minutes.

Mortar that falls to the scaffold or the ground should be discarded.

► **Laying the First Course**

The foundation or footing must be in place, and reinforcement rods in place, if required.

The first course of any block wall must be properly positioned to ensure the exact position of each block in the first course.

The rest of the wall is to be built on top of it, and adjustment to position cannot be made after the first course is embedded in mortar.

If more than one wall is to be made, the first course of all walls should be laid out together to test the exact position of all of the corner blocks.

<u>CAUTION</u>

When building a block wall, blocks must be kept dry.

Do not moisten blocks or permit them to become moistened during construction.

Cover incomplete wall sections and block supplies at the end of the work day to protect them from rain, dew or sprinklers.

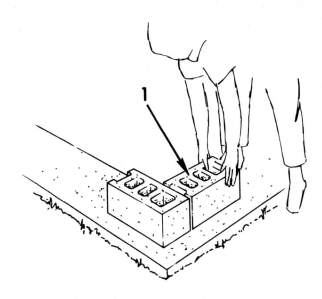

Concrete blocks are thicker on one side [1] than the other. Blocks must be placed with thicker side facing up.

1. Place the corner blocks at desired locations on foundation.

Laying the First Course

When placing the first course of stretcher blocks, keep in mind the positions of control joints, if required.

2. Place the first course of stretcher blocks [1] and [2] between the corner blocks. Allow an average of 3/8-inch between blocks for mortar joint. Strips of wood 3/8-inch wide can be used as a guide for spacing blocks.

After you are satisfied with the exact position of each block in the first course of all of the walls, continue.

3. Stretch a course string [3] to mark the top outer edge of the first course. Position the string about 1/4-inch away from the face of the wall to avoid striking it when laying the blocks.

4. Remove all blocks.

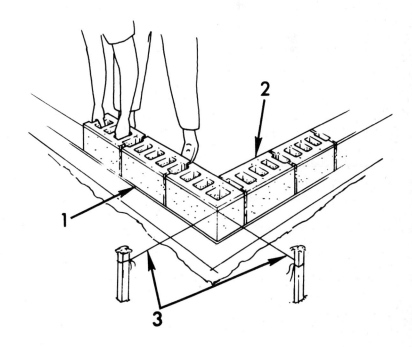

LAYING CONCRETE BLOCKS

Laying the First Course

In the next step, make mortar bed long enough to place three blocks, and as wide as the blocks.

5. Spread a bed of mortar [1] approximately 1 inch thick on foundation.

6. Make edges of mortar bed thicker than center of bed.

When placing blocks on the mortar bed, the thicker side must be facing up.

7. Place corner block at desired position. Tap block into mortar bed.

When placing the corner block, be sure that block is:

*   Properly aligned with the string.

*   Horizontally level.

*   Vertically plumb.

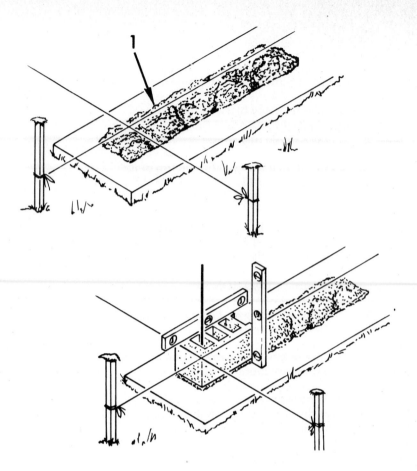

Laying the First Course

8. Place stretcher block on end. Spread a layer of mortar, approximately 3/4-inch thick, on edges of block.

9. Place stretcher block in mortar bed, with mortared end against the corner block. Tap stretcher block into mortar bed.

When placing stretcher blocks, be sure that blocks are:

*   Properly aligned with the string and each other.

*   Horizontally level.

*   Vertically plumb.

*   Vertical mortar joint [1] is about 3/8-inch wide.

10. Cut and remove excess mortar from vertical mortar joints.

11. Place the next block in the same manner.

Excess mortar may be spread on the edges of the next block, or remixed with the mortar on the mortar board.

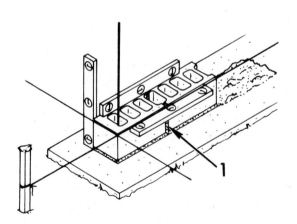

Laying the First Course

After three blocks have been mortared in place, repeat the procedures at the other end of the wall.

Do not permit the vertical mortar joint to be too wide, or the last block will not fit into the opening.

Mortar joints may vary from 1/4 to 5/8-inch. Mortar joints as wide as 3/4-inch at the last block are tolerable.

After three blocks have been mortared into place at the ends of the wall, the rest of the stretcher blocks of the first course may be installed.

When laying the first course of stretcher blocks, frequent and accurate checks are required to ensure that blocks are:

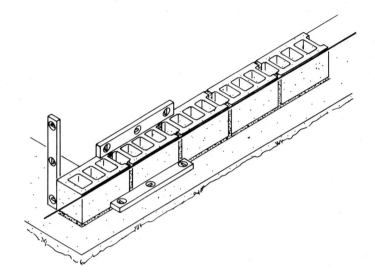

- Properly aligned with the string and each other.

- Horizontally level.

- Vertically plumb.

- Mortar joints are not too thick.

Laying the First Course

To install the last block in a course:

1. Spread a layer of mortar approximately 3/4-inch thick on both ends of the block [3].

2. Spread a layer of mortar approximately 3/4-inch thick on the ends of each block [1] and [4] in the course.

3. Spread a bed of mortar [2], the same as for other blocks in the course.

4. Carefully place block in opening. Tap block into place.

5. Check that block is even with string, horizontally level and vertically plumb.

6. Cut and remove excess mortar from mortar joints.

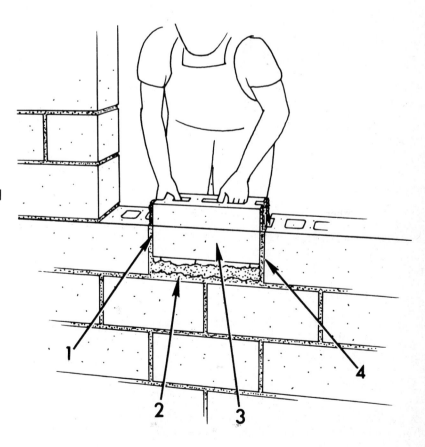

LAYING CONCRETE BLOCKS

▶ Building Up Corners and Ends

Walls can end in any of three ways:

- Free end [1].

- Joining at a corner [2].

- Joining at another wall [3].

Corners and ends are built up after the first course of each wall is mortared in place.

If you are building up the free end of a wall, go to next section (below).

If you are building up corners, go to Page 79.

If you are joining walls at another wall, go to Page 80.

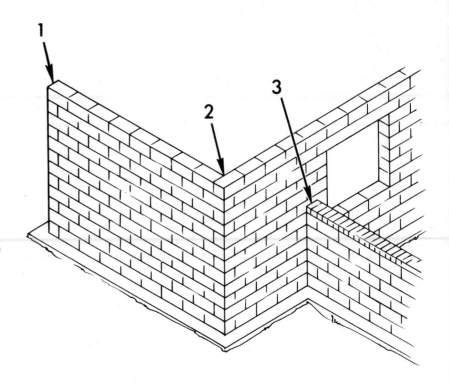

▶ Building Up Ends of Walls

End of walls may be built up several courses ahead of the stretcher courses.

The height of each end may be accurately checked by using a story board [1] to ensure that the top of each end is the same height as the other end.

Half blocks [2] are used in alternate courses to ensure that the vertical mortar joints are staggered.

The horizontal spacing of the vertical mortar joints may be checked by placing a straight-edge [3] diagonally across the corners of the blocks.

Check frequently that:

- The faces of all blocks are even with each other [4].

- The ends of all blocks are even with each other.

- Each course is horizontally level [5] and vertically plumb [6].

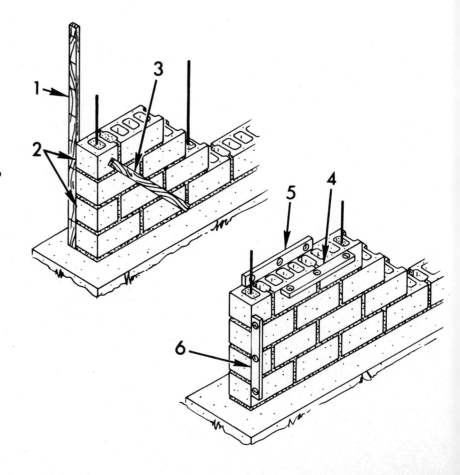

▶ Building Up Corners

The corner ends of both walls must be built up
concurrently.

Corner courses may be built up several courses
ahead of the stretcher courses.

The height of each corner course may be
accurately checked by using a story board [1]
to ensure that each corner course is the same
height as the other corner courses.

Full-size corner blocks [3] are used lengthwise
in alternate courses of each wall to ensure that
the vertical mortar joints of each wall are
staggered.

The horizontal spacing of the vertical mortar
joints may be checked by placing a straight-
edge [2] diagonally across the corners of
the blocks.

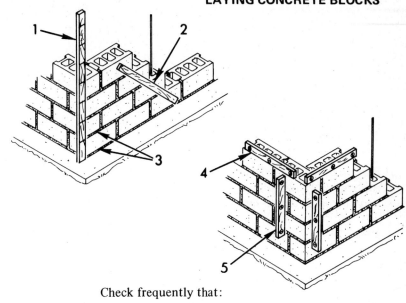

Check frequently that:

- The faces and ends of all blocks in each
  wall are even with each other.

- Each course in each wall is horizontally
  level [4] and vertically plumb [5].

Building Up Corners

If one or both of the walls are NOT load-bearing
walls, go to Page 80 (bottom).

If both of the walls are load-bearing walls, they
should be built up concurrently.

A steel strap [1] must be provided for attaching
the terminating wall to the continuous wall.

Steel straps must be embedded in the continuous wall at vertical intervals of not more
than 48 inches.

No mortar joint is required at the point where
the two walls meet. If you desire to fill the joint,
caulking compound may be used.

The straps have a 2-inch right-angle bend at each
end. The bends must be embedded in the core of
the blocks of each wall, and the cores filled with
grout.

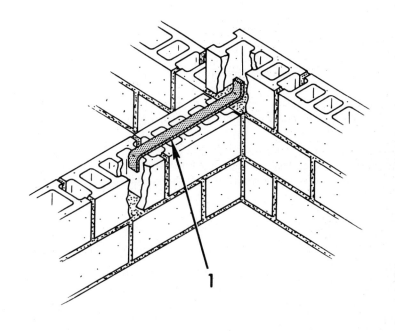

LAYING CONCRETE BLOCKS

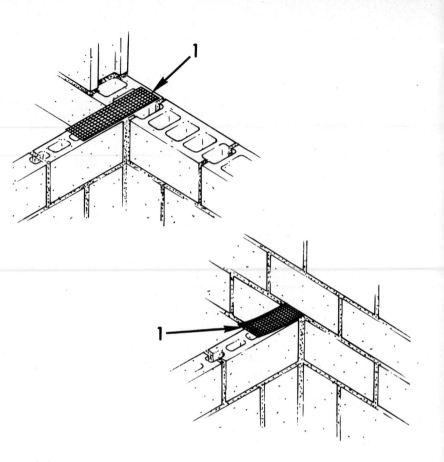

▶ Joining Walls

Provision must be made in the continuous wall for attaching the terminating wall to it.

Both walls may be built up concurrently or the continuous wall built up first and the terminating wall built up later.

Straps [1] of wire cloth or metal lath may be used for attaching the walls together.

Straps must be embedded in the continuous wall at vertical intervals of not more than 16 inches.

No mortar joint is required at the point where the two walls meet. If you desire to fill the joint, caulking compound may be used.

▶ Building Up Walls

After the first course is mortared in place and the corners are built up or started, the stretcher courses may be laid. Keep in mind the position of control joints, if required.

The procedure for laying stretcher courses is the same for all courses except the first course.

After each block is tapped into proper position, the excess mortar should be removed from the mortar joint.

The excess mortar may be spread on the edges of the next block or remixed with the mortar on the mortar board.

Blocks must be placed with thicker end facing up.

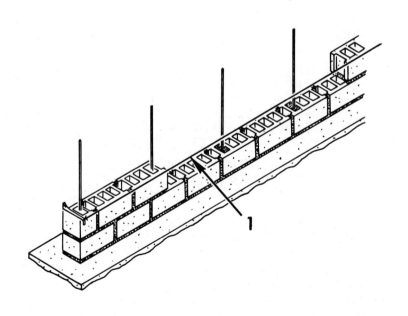

1. Stretch a string [1] between the built-up corners to mark the desired position of the top edge of the next course.

If the first course is level and the corner courses are level and plumb, the string will be level too, and used as your horizontal reference while laying the next course.

Building Up Walls

Normally, mortar is spread on the face shells of the block in the bottom course.

Sometimes the building codes require a full mortar bed. If a full mortar bed is required, mortar must be spread on the cross webs too.

2. Spread a bed of mortar [1] approximately 3/4-inch thick on bottom course.

3. Spread a layer of mortar approximately 3/4-inch thick on edges of block [2].

4. Place block in mortar bed with mortared edge [3] against previous block in the course. Tap block [4] into mortar bed.

When placing blocks, be sure that:

●   Vertical and horizontal mortar joints are approximately 3/8-inch thick.

●   Blocks are properly aligned with the course string and each other.

Building Up Walls

When you come to the last block in the course, it is installed the same way as the last block in the first course.

1. Spread a layer of mortar approximately 3/4-inch thick on both ends of block [3].

2. Spread a layer of mortar approximately 3/4-inch thick on the ends of each block [1] in the course.

3. Spread a bed of mortar approximately 3/4-inch thick on bottom course [2].

4. Carefully place block into opening. Tap block into place.

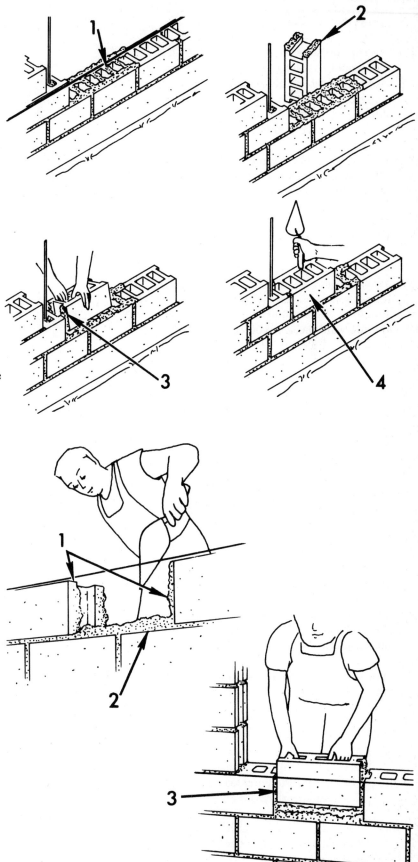

81

LAYING CONCRETE BLOCKS

▶ Laying the Top Course

The top course of blocks in a concrete block wall should be solid masonry to distribute the load uniformly and act as a barrier to insect or moisture penetration.

This may be accomplished in one of three ways:

- Solid top block is used to lay the top course [1].

- Cap block is used to lay the top course [2].

- Stretcher block is used and all of the cores are filled with grout [4].

The stretcher block method [4] is useful for foundation walls that are to support a wood frame structure. Anchor bolts [3] are placed in the top two courses at intervals of not more than 4 feet. Then the cores are filled with grout.

After the mortar and grout have hardened, a wooden sill may be fastened to the wall.

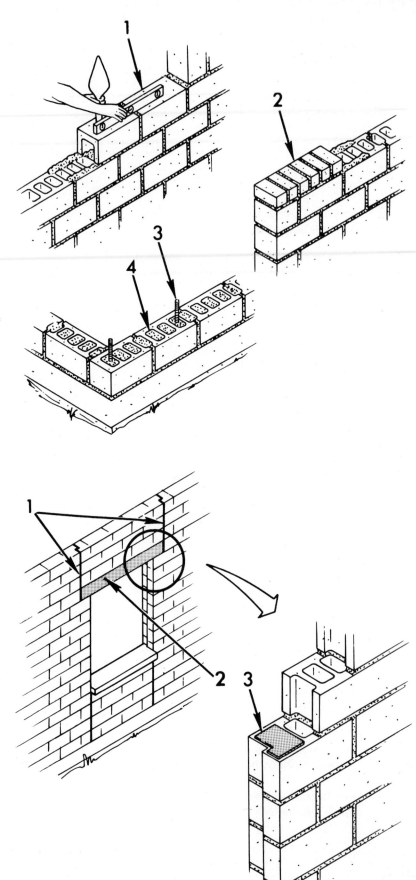

▶ Laying Lintels

The masonry structure that spans the top of an opening in a wall and carries the load above the opening is called a lintel [2].

Except when spanning large openings, a control joint [1] is needed on only one side of the opening.

For modular doors and windows, precast lintels are available, made of reinforced concrete.

Lintels may be made using stretcher blocks or header block. If making lintels, go to Page 83.

When installing precast lintels:

- A non-corroding steel plate [3] should be used, with a full mortar bed over the plate, to distribute the lintel load uniformly at the contact surface.

- Plate [3] should be on the control joint side of the opening if only one control joint is used.

Laying Lintels

- A non-corroding steel angle [1] must be used to span the opening and support the load.

- A full mortar bed must be used to distribute the lintel load uniformly at the contact surface [2].

When making a lintel using header blocks:

- A non-corroding steel angle [3] must be used to span the opening and support the load of the header blocks [4].

- A non-corroding steel angle [6] must be used to span the opening and support the header row of bricks [7].

- A full mortar bed must be used to distribute the lintel load uniformly at the contact surface [5].

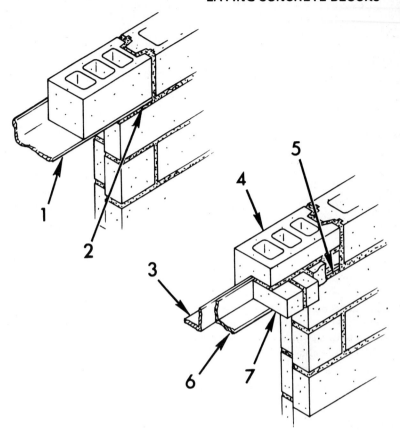

▶ **Constructing Control Joints**

To control cracks occurring in block walls caused by unusual stresses, control joints are built into the wall at the points where the stresses concentrate.

Control joints [1] are vertical seams, intended to permit the stresses to be relieved without breaking the mortar joints of the wall.

Control joints are useful:

- Above window and door openings.

- Where walls intersect at other than a corner.

- At intervals of not more than 20 feet in long continuous walls.

If control joints are exposed to weather or view, the seam may be filled with caulking compound.

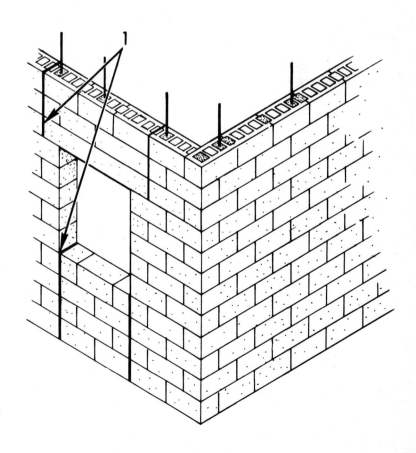

LAYING CONCRETE BLOCKS

Constructing Control Joints

There are three generally accepted methods of making control joints:

- Whole and half control joint blocks [1] are built into the wall to make a vertical seam.

  The control joint blocks provide lateral support to the wall sections on each side of the control joint by tongue-and-groove design.

- Whole and half jamb blocks [2] are built into the wall to make a vertical seam.

  Lateral support to the wall sections on each side of the control joints is accomplished by overlapping the jamb cutouts and imbedding a "Z" bar [3] in every other mortar bed across the control joint.

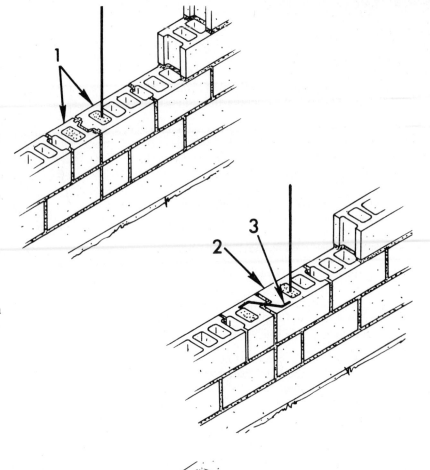

Constructing Control Joints

- Whole and half stretcher blocks [1] are built into the wall to make a vertical seam.

  Lateral support to the wall sections on each side of the control joint may be accomplished in one of two ways:

  A "Z" bar [2] imbedded in every other mortar bed across the control joint.

  Building paper [3] is inserted into the core hole and the core hole is grouted [4]. After the grout hardens, lateral support is provided by tongue-and-groove action.

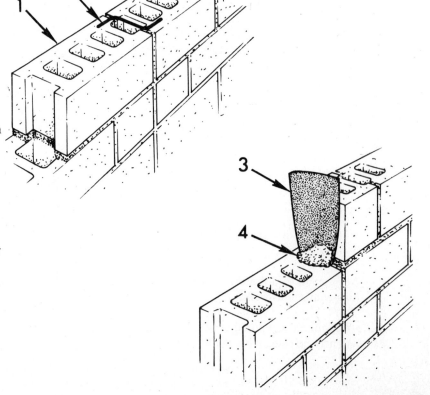

Constructing Control Joints

To keep control joints unnoticeable, the vertical mortar joints [1] should all be the same width and plumb.

After the mortar in the vertical seam becomes stiff, but before it hardens, mortar should be removed to a depth of 3/4-inch.

If the control joint is exposed to weather or view, the seam may be filled with caulking compound.

Caulking compounds may not adhere to the concrete blocks unless a primer coating is applied. Follow manufacturer's instructions when using caulking compounds and primers.

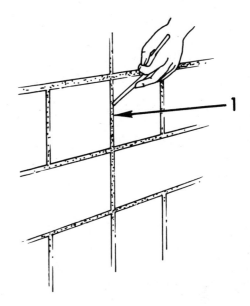

▶ Grouting

Grouting is the process of filling the cores of concrete blocks with mortar or grout.

Grouting is required when:

- Installing steel straps [2] in intersecting load-bearing walls.

- Installing anchor bolts [4].

- Making the top course [3] of foundation walls.

- Making reinforced walls [1].

Building codes will inform you of specific grouting requirements.

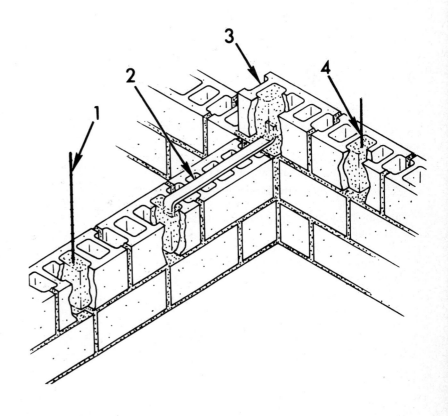

LAYING CONCRETE BLOCKS

Grouting

If core is to be completely filled from bottom to top, cores may be filled:

- As each course is completed.

- After 2 or 3 courses have been completed.

- After the wall is completed but before sills or caps are installed.

If the core is not to be completely filled from bottom to top, wire cloth [1] or metal lath may be embedded in the mortar bedding of the affected blocks, so that grout fills only the affected blocks.

You must know when you plan your block wall which cores must be filled with grout and where to use metal lath.

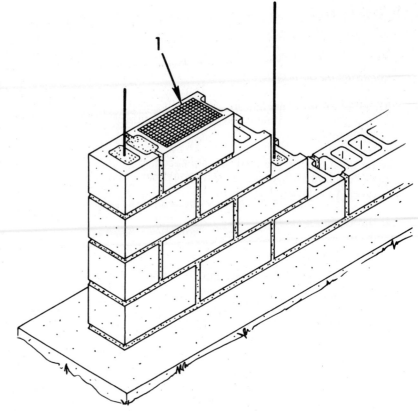

▶ Finishing Mortar Joints

Sometimes weathertight joints and neat appearance are desired in concrete block walls.

A mason's jointer [1] may be used to compact the mortar into the mortar joints to make joints weathertight and uniform in appearance.

Mortar should be compacted after it has become firm, before it hardens.

Horizontal mortar joints [2] should be compacted first, then the vertical joints [3].

After the joints have been compacted, any mortar burrs should be removed using the trowel [4] or by rubbing the surface with a heavy, coarse rag.

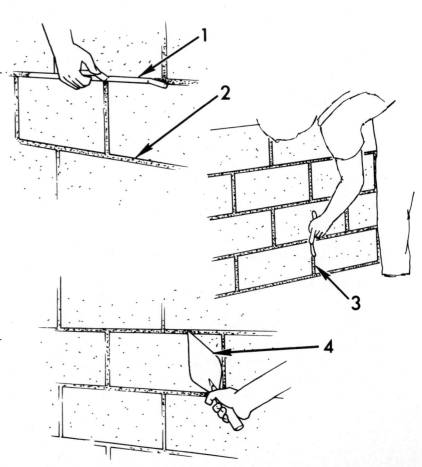

► Parging

Basement walls made of concrete blocks must be "weatherproofed" by parging [1].

Parging is a plaster-like coating of mortar, applied to the surface in two coats [2] and [3]. Each coat is approximately 1/4-inch thick.

The first parging coat is troweled firmly onto the wall surface after it has been wetted but not soaked.

Before the first coat hardens, it must be scratch-troweled to roughen the surface.

The first coat must be kept damp, but not soaked, while parging cures for 24 hours.

The second parging coat is troweled firmly over the first coat. The second coat should extend over the footing [4] to form a coving to protect the wall-to-footing joint.

The second coat must be kept damp, but not soaked, while parging cures for 48 hours.

Parging

If dampness may be a problem, as in poorly drained or wet soils, the parging may need to be coated with 2 coats of a hot bituminous sealing material or a good commercial waterproofing compound.

Parging coats must be thoroughly cured before applying bituminous material or waterproofing compounds.

Follow the manufacturer's instructions when applying bituminous material or waterproofing compounds.

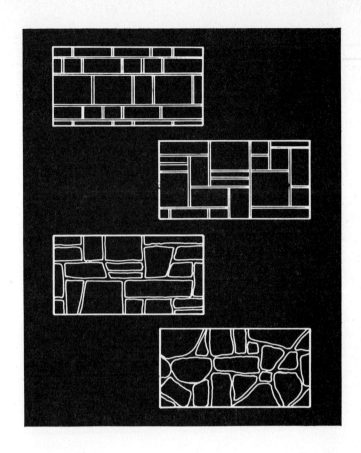

STONEWORK

Stone may be used as masonry material for many kinds of decorative walls and paving. Stone can be used to achieve a more casual appearance than can be achieved using bricks.

Because stonework will have a more random grouping of sizes and shapes, stonework structures can not be planned with the same precision as other masonry structures.

When planning your stonework addition, consider future architectural and landscape changes. Make provisions for these changes at this time.

Some future changes to consider might be:

- Water provisions for hose connections or sprinkler systems.

- Electrical provisions for future utilities or lighting.

- Gas line provisions for future outdoor barbeque, pool heater, etc.

Consider future growth of landscape vegetation and their root systems. You may need to re-locate or remove a tree or shrub.

Start your stonework addition by making a scale drawing. The drawing can help you determine the amount of materials you will need.

Take your drawing to your masonry supplier to work out the amount, kind and cost of materials required for your addition.

When buying stones for a wall, steps or a fireplace, get a variety of sizes and shapes. This will avoid the need for too much mortar in the mortar joints. Mortar tends to shrink as it hardens. If mortar joints are too thick, cracks will appear in the joints.

An alternative to purchasing masonry material is to find what materials are available at no cost. Stones that are native to your area will look most natural in your stonework.

Natural stones must not crumble or break to be used for your stonework. They must have sufficient strength and durability for their intended use.

TOOLS AND SUPPLIES

The tools and supplies required for stonework are the same as those required for brickwork. Page 10.

In addition, the following tools are required for making stone paving:

- Power drill [1] and masonry disc [2] for trimming paving stones to fit.
- Rubber mallet [3] for tapping paving stones level during installation.
- Slate cutter [4] if using slate paving stones.

Mortar used for stonework is the same as that used for brickwork. See Mixing Mortar, Page 13.

The following mortar mixture should be used for stonework. Proportions are measured by volume:

1 part white Portland cement. Grey Portland cement can cause stains and should not be used.

3 parts clean, damp mortar sand.

1/4 part hydrated lime.

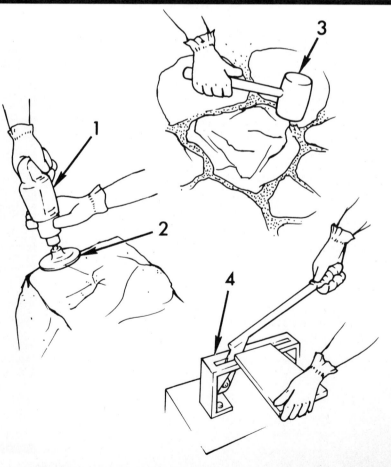

The design of a stone wall depends largely on your own personal preferences. There is a variety of shapes, sizes and textures of stone building material.

There are four basic types of stone wall design:

Coursed ashlar walls [1]

Random ashlar walls [2]

Coursed rubble walls [3]

Random rubble walls [4]

Ashlar walls [1] and [2] are made of regular-shaped stones [5]. For coursed walls [1], the stones are laid in continuous horizontal rows. They can be built according to the procedures in the BRICKWORK section. Random walls [2] are built with a more random pattern.

Rubble walls [3] and [4] are made of rough-shaped stones [6] and [7]. For coursed rubble walls [3], the stones [6] are similar in size and are laid in essentially horizontal rows. Random walls [4] are built with irregular-shaped stones [7] in a random pattern.

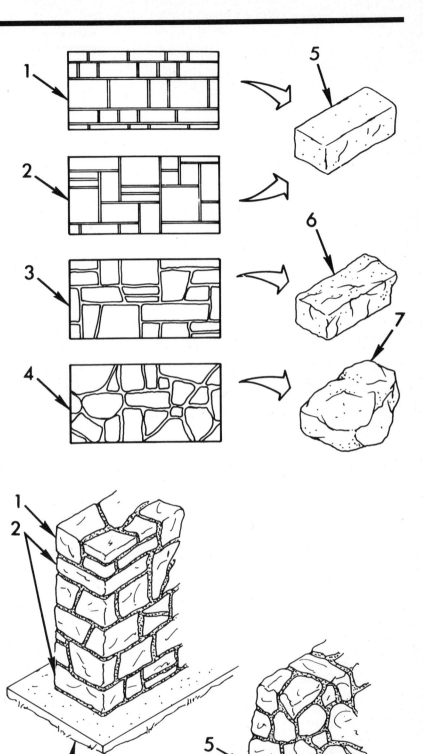

Stone walls must be built on a footing [3] or foundation slab. Because stone walls are not structural walls, it is not necessary to build the footing exactly level. See the POURED CONCRETE section for instructions on building a footing or slab for your wall.

Use of bonding stones [2, 5] is essential in stone walls. Bonding stones are stones which extend through the entire width of the wall, helping to stabilize the structure. Plan to use a bonding stone for every 6- to 10-square feet of wall surface.

Rubble walls [4] should be built thicker at the base than at the top. Walls should slope about 1 inch for each foot in height. Ashlar walls [1] do not require a slope.

When building a stone wall, carefully plan the
position of the mortar joints [1]. Vertical mortar
joints must be offset from those below. The
vertical joint extending through the wall must
also be offset.

Fill large gaps between stones with smaller
stones [2]. Do not use large quantities of mortar
to fill a large gap.

Avoid leaving cavities in mortar joints [1] where
water can collect. Moisture trapped in mortar
joints can damage the wall in areas subject to
freezing. Compress the mortar joints as soon
as the mortar begins to harden.

Before beginning to make your wall, be sure
that all stones are clean. Using clean stones
insures a good mortar bond. As a visual reference for building straight ends of the wall,
temporary stakes [3] can be used.

The procedure for making stone walls begins
below.

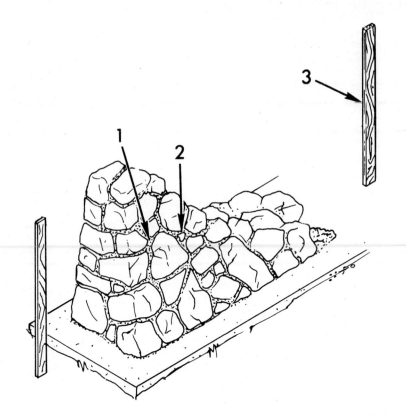

1. Place bonding stones [1] at each end of
    the wall and at appropriate intervals along
    the first course.

2. Place the first course of stones [3] in
    position between the bonding stones [1].
    Place stones with their flat side down.

3. Adjust the size, shape and position of the
    stones [3] to minimize large gaps. Use
    small stones [4] to fill any large gaps.

When you are satisfied with the position of each
stone, continue.

4. Remove several stones. Spread a bed of
    mortar about 2 inches thick on the foot-
    ing [2] or base.

5. Place the stones back in their original
    position in the bed of mortar.

6. Repeat Steps 4 and 5 until all stones in
    the first course are mortared in position.

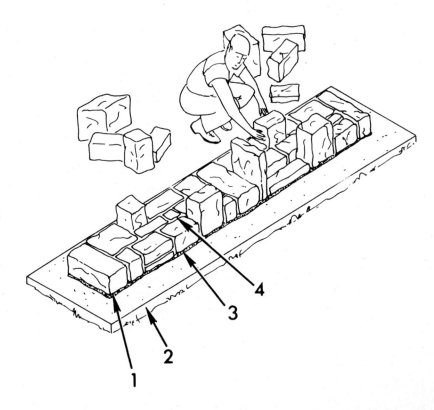

The ends of the wall are built up after the first course of stones is in position. Stones may be cut to fit any position desired as you make the wall.

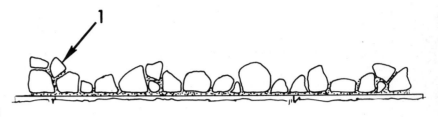

7. Place several stones [1] at one end of the wall. Adjust their size, shape and position until satisfied with their appearance. Use bonding stones where required.

8. Remove the stones [1]. Spread a bed of mortar on the stones below.

9. Place the stones [1] back in their original position. Press them firmly into the mortar.

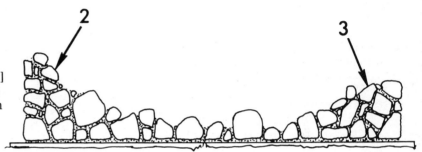

10. Repeat Steps 7-10 to build up both ends [2] and [3] of the wall. Check frequently that the mortar is making complete contact with each stone and that the alignment of the wall is correct.

When both ends of the wall are built up, the remaining courses of stone are laid. Stones may be cut to fit any position desired.

11. Place a course of stones [1] between the ends of the wall.

12. Adjust the size, shape and position of the stones [1] until you are satisfied with their appearance. Use bonding stones where required.

13. Remove the stones [1]. Spread a bed of mortar on the stones below.

14. Place the stones [1] back in their original position. Press them firmly into the mortar.

15. Repeat Steps 11-15 to finish building the wall. Check frequently that the mortar is making complete contact with each stone and that the alignment of the wall is correct.

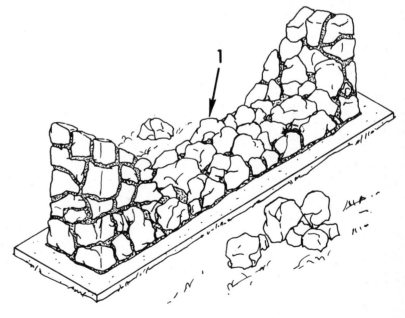

The types of stone commonly used to build paving are flagstone and slate. These materials are hard and flat, making them an excellent choice for decorative paving.

The design of stone paving and its building procedures are similar to those for brick paving. Stones, like bricks, can be laid on a concrete slab or in a bed of sand. If laying stones on a concrete slab, mortar or a commercial mastic adhesive can be used.

When building stone paving, it is necessary to cut and fit each stone [1] before placing it in its permanent position.

Follow the instructions in the BRICKWORK section to make stone paving.

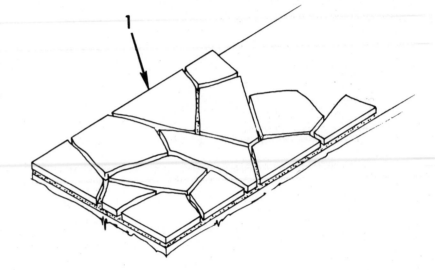

PROJECT	DATE	MATERIAL

PROJECT	DATE	MATERIAL

PROJECT	DATE	MATERIAL

PROJECT	DATE	MATERIAL

NOTES

NOTES

NOTES

NOTES